International Congress of Mechanical Engineering and Agricultural Sciences

(CIIMCA 2013)

IOP Conference Series: Materials Science and Engineering Volume 59

Santander, Colombia
8 – 11 October 2013

ISBN: 978-1-63439-870-1
ISSN: 1757-8981

Printed from e-media with permission by:

Curran Associates, Inc.
57 Morehouse Lane
Red Hook, NY 12571

CURRAN ASSOCIATES INC.
proceedings
.com

Some format issues inherent in the e-media version may also appear in this print version.

Printed by Curran Associates, Inc. (2015)

Institute of Physics
Dirac House, Temple Back
Bristol BS1 6BE UK

Phone: 44 1 17 929 7481
Fax: 44 1 17 920 0979

techtracking@iop.org

Additional copies of this publication are available from:

Curran Associates, Inc.
57 Morehouse Lane
Red Hook, NY 12571 USA
Phone: 845-758-0400
Fax: 845-758-2634
Email: curran@proceedings.com
Web: www.proceedings.com

TABLE OF CONTENTS

IOP Conf. Series: Materials Science and Engineering **59** (2014) 011001 doi:10.1088/1757-899X/59/1/011001

Preface

The organizing committee of The International Congress of Mechanical Engineering and Agricultural Sciences – CIIMCA 2013 – are pleased to present CIIMCA-2013: the first international conference focused on subjects of materials science, mechanical engineering and renewable energy organized by Mechanical Engineering Faculty of the "Universidad Pontificia Bolivariana" in Bucaramanga, Colombia. This conference aims to be a place to produce discussions on whole topics of the congress, between the scientists of Colombia and the world.

We strongly believe that knowledge is fundamental to the development of our countries. For that reason this multidisciplinary conference is looking forward to integrate engineering, agricultural science and nanoscience and nanotechnology to produce a synergy of this area of knowledge and to achieve scientific and technological developments. Agriculture is a very important topic for our conference; in Colombia, agricultural science needs more attention from the scientific community and the government. In the Faculty of Mechanical Engineering we are beginning to work on these issues to produce knowledge and improve the conditions in our country.

The CIIMCA conference is a great opportunity to create interpersonal relationships and networks between scientists around the world. The interaction between scientists is very important in the process of the construction of knowledge. The general chairman encourages and invites you to make friends, relationships and participate strongly in the symposia and all program activities.

PhD Aduljay Remolina-Millán
Principal Chairman, International Mechanical Engineering and Agricultural Sciences Congress – CIIMCA
Msc Emil Hernández-Arroyo
Principal Chairman, International Mechanical Engineering and Agricultural Sciences Congress – CIIMCA

IOP Conf. Series: Materials Science and Engineering **59** (2014) 011001 doi:10.1088/1757-899X/59/1/011001

Special Guests and Organizing Committee. From left to right. Sitting: Jorge Díaz, Unipamplona, Colombia; Dra Nayibe Guerrero, CINVESTAV; México; Dr Giovanni Ramírez, ANL, USA; Jhoston Zamora, Vortex company, Colombia; Dr José Rivas, INL, Spain–Portugal; Dra Sandra Rodil, IIM-UNAM, Mexico; Rafael Parúas, Cubasolar, Cuba; Fabian Carreño, QES, Germany. Standing: William Sánchez, UPB; Julian Carrillo, UPB; Diego, UPB; Emil Hernández, UPB; Marta Lucía, UPB; Katy, UPB; Marlon López, UPB; Ingrid, UPB; Elías Villdiego, UPB; Aduljay Remolina, UPB and Luis Solano, UPB.

IOP Conf. Series: Materials Science and Engineering **59** (2014) 011001 doi:10.1088/1757-899X/59/1/011001

"Universidad Pontificia Bolivariana seccional Bucaramanga" host of the first International Mechanical Engineering and Agricultural Sciences Congress – CIIMCA 2013 – Floridablanaca, Colombia.

Closure of CIIMCA 2013.

IOP Conf. Series: Materials Science and Engineering **59** (2014) 011001 doi:10.1088/1757-899X/59/1/011001

Editorial Committee

IOP Conf. Series: Materials Science and Engineering **59** (2014) 011001 doi:10.1088/1757-899X/59/1/011001

Universidad Nacional de Colombia, Bogotá, Colombia

PhD Mauricio Moreno Téllez
Carlosmauricio.moreno@uptc.edu.co
Universidad Pedagógica y Tecnológica de Colombia (UPTC), Tunja, Colombia

PhD José Luis Tristancho Reyes
josetris@utp.edu.co
Universidad Tecnológica de Pereira, Pereira, Colombia

PhD Diana Marulanda
dmaritzamc@gmail.com
Universidad Antonio Nariño Bogotá, Colombia

PhD Ariel René Carreño
rene.carreno@upb.edu.co
Universidad Pontificia Bolivariana seccional Bucaramanga, Floridablanca, Colombia

PhD Omar Pinzón Ardila
omar.pinzon@upb.edu.co
Universidad Pontificia Bolivariana seccional Bucaramanga, Floridablanca, Colombia

PhD Yolanda Torres
yolanda.torres@usantoto.edu.com
Universidad Santo Tomás, seccional Tunja, Tunja, Colombia

PhD Edwin Y Gómez Pachón
edwin.gomez02@uptc.edu.co, edwinyesidgom@yahoo.com.mx
Universidad Pedagógica y Tecnológica de Colombia-UPTC, seccional Duitama, Duitama, Colombia

PhD Aldo Pardo García
apardo13@hotmail.com
Universidad de Pamplona, Pamplona, Colombia

MSc Ely Dannier Valbuena Niño
Ely.valubuena@upb.edu.co
Universidad Federal de Rio de Janeiro, Rio de Janeiro, Brasil

MSc Jorge Luis Díaz
jdiazcu@gmail.com
Universidad de Pamplona, Pamplona, Colombia

MSc Carlos Jaimes
carlos.jaimes@upb.edu.co
Universidad Pontificia Bolivariana seccional Bucaramanga, Floridablanca, Colombia

IOP Conf. Series: Materials Science and Engineering **59** (2014) 011001 doi:10.1088/1757-899X/59/1/011001

MSc Javier Castellanos
javier.castellanos@upb.edu.co
Universidad Pontificia Bolivariana seccional Bucaramanga, Floridablanca, Colombia

MSc Emil Hernandez Arroyo
emil.hernandez@upb.edu.co
Universidad Pontificia Bolivariana seccional Bucaramanga, Floridablanca, Colombia

PhD Aduljay Remolina Millán
aduljay.remolina@upb.edu.co aduljayr@yahoo.com
Universidad Pontificia Bolivariana seccional Bucaramanga, Floridablanca, Colombia

Acknowledgments

The organizing committee wants to thank Anete Ashton, Conference Commissioning Editor of the IOP Conference Series for her support and help in the process of publishing the papers from CIIMCA 2013 in the IOP Conference Series.

The organizing committee wants to thank specially the Editorial Committee for their great support, participation and valuable contribution to CIIMCA 2013.

The organizing committee wants to thank specially the Mechanical Engineering Faculty and the "Universidad Pontificia Bolivariana seccional Bucaramanga" for its great support and funds given to development of CIIMCA 2013.

The organizing committee wants to thank specially the authors for their great contribution to the development of Colombia.

IOP Conf. Series: Materials Science and Engineering **59** (2014) 011002 doi:10.1088/1757-899X/59/1/011002

Peer review statement

All papers published in this volume of *IOP Conference Series: Materials Science and Engineering* have been peer reviewed through processes administered by the proceedings Editors. Reviews were conducted by expert referees to the professional and scientific standards expected of a proceedings journal published by IOP Publishing.

The International Congress of Mechanical Engineering and Agricultural Sciences 2013 IOP Publishing
IOP Conf. Series: Materials Science and Engineering **59** (2014) 012001 doi:10.1088/1757-899X/59/1/012001

Synthesis and characterization of inorganic polymers from the alkali activation of an aluminosilicate.

C P González [1]*, A M Montaño[1,2], A K González[2], C A Ríos[3].
[1] Grupo de Investigación en Materiales "GIM", Departamento de Ciencias Básicas, Universidad Pontifica Bolivariana, Seccional Bucaramanga. Bucaramanga. Colombia.
[2] Escuela de Química, Universidad Industrial de Santander. Bucaramanga. Colombia.
[3]Grupo de Investigación en Geología Básica y Aplicada (GIGBA). Escuela de Geología. Universidad Industrial de Santander, Bucaramanga, Colombia

E-mail: claudia.gonzalez@upb.edu.co

Abstract. This paper presents the results of the synthesis and characterization of inorganic polymers (IP) from aluminosilicates: bentonite (BT) and pumice (PP). The synthesis of IP, was carried out by two methods involving alkaline activation, at room temperature and 80 ± 5 °C, using as activating agent sodium silicate both commercial and analytical (Na_2SiO_3). Sodium hydroxide (NaOH) at 3 M, 7 M and 12 M was added. A lower degree of polymerization was obtained by using analytical precursors subjected to room temperature and 80 ± 5°C. Replacement of heating by the use of the commercial activating agent with greater alkalinity allows the formation of a 3D network. The materials were structurally characterized by FTIR spectroscopy with Attenuated Reflectance (ATR), Scanning Electron Microscope (SEM) and X -ray diffraction (DRX).

1. Introduction.

In the last decades research has been conducted to improve mechanical, electrical properties and durability of Portland Cement. Geopolymers combining properties of polymers and ceramics which are obtained from chemical reaction called geopolymerization. This reaction is carried out in three steps:(i) dissolution in alkaline solution, (ii) reorganization and diffusion of dissolved ions with formation of small coagulated structures and (iii) polycondensation of soluble species to form hydrated products. On that process the raw material -aluminosilicate compounds of silicon and aluminum oxide- are activated by an oxidizing agent [1-3].
Results of the synthesis of two polymers or inorganic geopolymer (IP) by two methods, using two sources of aluminosilicates, are presented in this work: Bentonite (BT) and Pumice (PP) modified and activated at two temperatures (room temperature and 80 ± 5°C). The samples were characterized structurally by experimental techniques: Attenuated Total Reflectance Infrared Spectroscopy (ATR-FTIR), Scanning Electron microscopy (SEM) and X-ray Diffraction (XRD)

2. Experimental.

As aluminosilicates source two clays correspond to bentonite (BT) obtained from Chemical Campota, Bogota (Colombia) and pumice (PP) extracted from Sabaneta, Antioquia (Colombia) are used. IP synthesis was performed by alkaline activation at room temperature and at 80 ± 5 °C in metallic containers of 1L capacity. Activating agent was sodium silicate (Na_2SiO_3) both, commercial and analytical degree, and sodium hydroxide (NaOH) 3, 7 and 12 M in 2:1 proportion. Sand 1:1 was also

[1]* To whom any correspondence should be addressed.

added to clay. The reaction mixture was left during 31 hours with vigorous stirring. Final product was dried at room temperature for 6 hours to obtain a hard paste which was ground to 4.77 mm of particle size according to ASTM C94 /C 94 M -09) [4]. The raw materials and the as-synthesized geopolymers were analyzed by Attenuated Total Reflectance Fourier Transformed Infrared Spectroscopy (ATR-FTIR Bruker Tensor 27). SEM images were obtained by field emission gun environmental scanning electron microscopy (FEI Quanta 650 FEG SEM), under the following analytical conditions: magnification = 2000-5000x, WD = 9.2-10.0 mm, HV = 5-20 kV, signal = Z Cont detector = ETD. X-Ray Diffraction (Bruker D8 ADVANCE diffractometer system), operating in DaVinci geometry with CuKα1 radiation at 40 kV and 30 mA.

3. Results and discussion

3.1. Geopolymerization reaction

Geopolymers bentonite (BT IP) and pumice (PP IP) were synthesized by two different methods (1 and 2) from the BT and PP aluminosilicates respectively, which by its lamellar structure guide the reaction geopolymerization in a plane by increasing the speed of the same [5]. The method 1 involves two temperatures between 20 -100 °C and the use of an activating aqueous medium provided by sodium silicate (Na_2SiO_3) of commercial grade for rapid polymerization of the aluminosilicate employed [1]. Initially this synthesis was performed at two temperatures, room temperature and to 80 ± 5 °C, the first non-disbanded the raw material in the activating solution, while the second disbanded precursors to later present the restructuring and polycondensation thereof for formation of the polymeric matrix.

In method 2, a different activating solution was used: equimolar ratio 2:1 silicate and analytical sodium hydroxide (Na2SiO3/NaOH) at 3, 7 and 12 M, with the purpose of to evaluate the effect on geopolymerization by increasing alkalinity in the reaction mixture [6]. This method, however, does not work well at any temperature. The dissolution of the raw materials was not given in the activating solution, not even by increasing the concentration of NaOH and by raising the relationship of SiO2/Na2O at the reaction medium[7]. So that synthesis of BT IP and PP IP was possible at 80 ± 5 °C and commercial Na_2SiO_3 higher alkalinity only.

3.2 Structural analysis

3.2.1 Analysis by attenuated reflectance infrared spectroscopy (ATR)

Precursors (BT and PP) and synthesized polymers by FTIR-ATR were analyzed to further the formation of the polymer matrix. The range of 700 – 2200 cm^{-1} (Figure 1)is remarkable because of the most important characteristics are evident: vibration of the Si-O and Al-O to 996 cm^{-1} for BT and 1018 cm^{-1} for PP; the presence of mineral oxides for both aluminosilicates materials in the region of 700-800 cm^{-1} is also found. Differences mainly being observed between the precursor (BT and PP) and BT IP and PP IP in a shift to higher wave numbers in the band corresponding to the vibrations Al-O and Si-O, attributed to the replacement of Al in the original structure of the skeleton of Si-O-Si precursor [8]. Also weak band at 1400-1600 cm^{-1} is related to stretching vibrations of terminal Si–OH groups in partially hydrated PI of BT and PP. From these last two bands IP formation through structural rearrangement of aluminosilicate sources (BT and PP) by alkaline solution is confirmed.

In addition, at FTIR- ATR spectra for synthesized IP, a displacement to shorter wavelengths for the mineral oxides bands is observed. This fact is due to the modification of aluminosilicates (Al_2O_3. $2SiO_2$) by the activating agent (Na_2SiO_3), in order to obtain both ionic species: aluminate (Al $(OH)^{4-}$) and monosilicate ($OSi(OH)^{3-}$), to form the corresponding polycondensation monomer ($n(OH)_3$-Si-O-Al-$(OH)_3$.

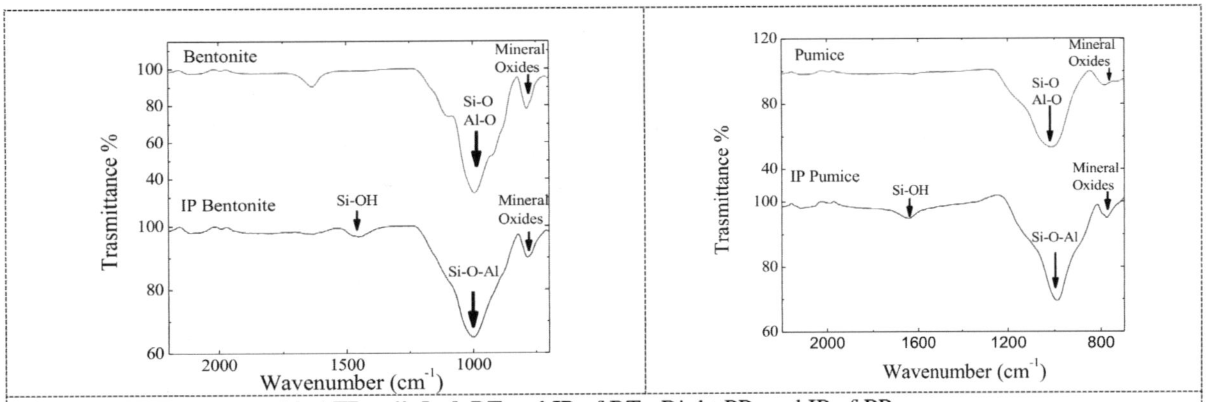

Figure 1. FTIR spectra with ATR cell: Left BT and IP of BT. Right PP, and IP of PP.

3.2.2 Analysis by Scanning Electron Microscope (SEM)

The SEM of the precursor materials and the synthesized IP micrographs are presented in Figure 2. For both precursors homogeneous morphology and similar size are observed BT (Figure 2a) and PP (Figure 2c). Obtained IP exhibits a very different conformation with islands of raw materials unpolymerized, (BT IP (Figure 2b) and PP IP (Figure 2d)) immersed in a polymeric matrix. The aluminosilicates is shown unreacted (white arrows) [9]. Besides the particle size is clearly smaller for IP obtained from both clays than for BT and PP, because of compaction of materials in a cementitious mixture with better molecular organization which brings on crystallinity to the geopolymer.

Figure 2. SEM micrographs of (a)BT (b)BT IP (c) PP (d)PP IP

3.2.3 Analysis by X-Ray Diffraction

The X-ray diffraction data for precursors (BT and PP) and the synthesized IP (BT IP and PP IP) are presented in Figure 3. Main mineral phases for BT are quartz, albite, gypsum, calcite, anorthoclase, kaolinite and montmorillonite. By alkaline activation orthoclase and muscovite appears in BT IP synthesized except by gypsum and anorthoclase [10]. Similarly for PP mineralogical phases corresponding to quartz, illite, albite, anorthite and iron oxide are found, keeping quartz and albite phases for PP IP synthesized. Cristobalite, muscovite and anorthoclase phases are observed also by activation alkaline.

It is also important to highlight the increase of intensity and resolution of the peaks for both IP, particularly in the quartz and anorthoclase phases, compared to those of precursors, due to the crystallinity achieved by geopolymerization, as it was observed by SEM.

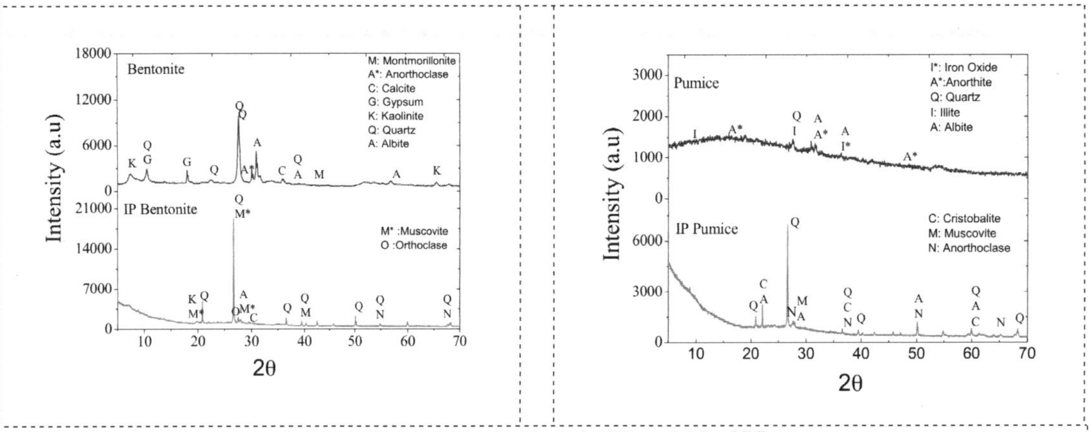

Figure 3. Diffractogram for: Left BT and PI of BT; right PP and PI of PP.

Conclusions

The main findings of this study can be summarized as follows:

- Geopolymerization reaction was successful at high temperature and in a highly alkaline medium. On that conditions polymerization was reach at short times.

-Structural characterization by FTIR-ATR, SEM and X-Ray Diffraction of IP synthesized, shows the structural rearrangement of the aluminosilicate sources (BT and PP) with improvement in crystallinity degree. It is also found that not all the raw material is polymerized suggesting that less quantities of clay could be used.

Acknowledgements

The authors express their gratitude to the School of Chemistry, X-ray diffraction Laboratory and Microscopy Laboratory of the Universidad Industrial de Santander and the Universidad Pontificia Bolivariana-Bucaramanga, for the financial support of the project 011-0610-9560.

References.

[1] Davidovits J 1991 J. Therm. **37** 1633.

[2]Lecomte I, Liégeois M, Rulmont A and Cloots R 2003 *J Mater Res* **18** 2571-2579.

[3] Zang Z, Wang H, Provis J, Bullen F, Reid A and Zhu Y 2012 *Therm Acta* **539** 23-33.

[4] ASTM C94 /C 94M – 09: Standard Specification for Ready-Mixed Concrete. *ASTM – American Society for Testing and Materials.*

[5] Wang H, LI H and YAN F 2005 *Colloids Surf* **268** 1–6.

[6] Pacheco Torgal F, Castro Gomes J and Jalali S 2008 *Const and Build Materl* **22** 1315–1322

[7] Lancellotti I, Catauro M, Ponzoni C, Bollino F, Leonelli C 2013 *J of Solid State Chem* **200** 341-348

[8] Rovnanik P 2010 *Const and Build Materl* **24** 1176–1183.

[9] Liew Y.M, Kamarudin H, Mustafa Al Bakri A. M, Lugman M, Khairul Nizar I, Ruzaidi M and Heah C.Y 2012 *Const and Build Materl* **30** 794–802.

[10] Zhang Z, Yao X and Zhu H 2010 *Appl Clay Sci* **49** 7–12.

IOP Conf. Series: Materials Science and Engineering **59** (2014) 012002 doi:10.1088/1757-899X/59/1/012002

Study of Metal Organic Chemical Vapour Deposition (MOCVD) semiconductors III-V hyperstructures with Secondary Ion Mass Spectrometry (SIMS).

J J Laserna[1]*, H Téllez Lozano[2], José M Vadillo[3] and D Padilla[4]

[1-3] Universidad de Málaga, Departamento de Química Analítica, Facultad de Ciencias, Málaga, España – Spain.
[2] International Institute for Carbon-Neutral Energy Research (WPI-I2CNER), Kyushu University, Japan.
[3] Escuela de Física, Universidad Industrial de Santander, Bucaramanga, Colombia.

E-mail: laserna@uma.es

Abstract. One of the most promising technologies in high efficiency solar cells is based on quaternary structures grown by epitaxial techniques as Metal Organic Chemical Vapour deposition (MOCVD). The semiconductors III-V structures are elaborated under tailored parameters, allowing the use of a broader area of the solar spectrum. Analytical techniques capable of providing accurate and precise information in cross sections about the composition and thickness of the layers are demanded. Secondary Ion Mass Spectrometry (SIMS) has been used for characterization of these structures due to its high depth resolution and sensitivity, stability and reproducibility. It was detected the diffusion process of Al and In across the cell interfaces and the layer diffusion over GaAs substrates. The Al diffusion was associated at incorrect incorporation of elements during growth process and the layer diffusion was associated at changes of manufacturing parameters. Such studies show the SIMS ability to diagnose of faults during the growth process, detection of impurities and incorrect diffusion of dopants that may affect the layer properties and the structure functionality.

1. Introduction

In the cell solar industry, silicon solar modules account a high portion of the market despite its moderate efficiency in the use of the whole solar spectrum. Due to this fact, a concentration device is commonly incorporated to improve its conversion efficiency [1-3]. However, it may cause the increase of the cell temperature with the solar exposition and the reduction in cell efficiency. Since 1980s, solar cells based in structures composed by layer of III-V semiconductors materials have been studied. They exhibit a direct band gap with tuneable absorption zone and allow the efficient use of the whole solar spectrum [4]. Although III-V solar cells have been considered too expensive because of the raw materials and processes involved in their production; the combination of these structures with the PV technology assures high efficiency with respect to conventional silicon solar cells [1]. Besides, the use of Ge substrates allows the manufacturing of thinner and lighter structures. These characteristics have made these solar cells the choice for space applications as the higher power–weight ratio permits to reduce weight and consequently reduce launch costs [5]

[1]* To whom any correspondence should be addressed.

The manufacturing solar cell industry requires analytical tools to study the fabrication process and properties in cross sections at high resolution, check the interface regions and detect diffusion process between substructures or incorrect growing process of layers that can degrade the solar cell efficiency [2-7]. SIMS has shown to be a sensitive technique for studying the distribution of dopants and contaminants over these materials, due to the complete atomic/molecular information provided and its excellent sensitivity and reproducibility. The cited characteristics allow to determine manufacturing defects or diffusion processes that can be affect its functionality or the physical layer properties [2,3,6]. In this paper, a study of the diffusion process in III-V semiconductor structures is presented, SIMS has been used for to characterize and study the structures become a diagnostic tool for elaboration process.

2. Experimental Setup

A solar cell structure (GaInP/InGaAs/Ga(In)As/Ge) consists of three subcells GaInP/Ga(In)As/Ge connected with tunnel junction (TJ) and are grown on a 140 μm thick Ge wafer by MOCVD. An operative cell shows specific values of structural, electrical and optical parameters that allow to obtain the best performance; defective cells don't show it. The efficiency of the cell is degraded by the surface defects, diffusion across the interfaces or misfit dislocations [2-7]. SIMS has been used to study and characterize a complete and operative structure (GaInP/InGaAs/Ga(In)As/Ge) with Ar ions [2], the analysis conditions were optimized to obtain good depth-resolution without compromising either the sensitivity or the reproducibility of the analysis. In this paper, a fully-operative cell (sample A) and a defective one (sample B) were studied under identical sputtering conditions as reported [2] but using O_2^+ to increment the sputtering rate. It has been studied if there are differences in layer composition or layer distribution of the samples that can be associated at its functionality. The signal SIMS corresponding to $^{27}Al^+$, $^{28}Si^+$, $^{31}P^+$, $^{64}Zn^+$, $^{69}Ga^+$, $^{75}As^+$, $^{115}In^+$ and $^{74}Ge^+$, elements presents in the structure were monitored.

In other hand, It was analysed the influence in the layer structure of the gas flow of the organic compounds in the MOCVD reactor during growth process. It was used a set of simple structures grown over GaAs substrate. The samples were elaborated keeping the parameters similar at the cells manufacturing parameters, except Ga flow. In the Table 1 is shown the gas flow values of the different elements in the growth process of samples. The signal SIMS corresponding to $^{27}Al^+$, $^{31}P^+$, $^{69}Ga^+$, $^{75}As^+$ and $^{115}In^+$ were monitored

Table 1. Flow values in the MOCVD reactor for the set of samples S1, S2, S3, S4, S5. The unities are sccm.

	Sample				
Element	1	2	3	4	5
Al	52.00	52.00	52.00	52.00	52.00
Ga	44.00	49.50	44.68	44.36	45.00
In	440.00	440.00	440.00	440.00	440.00
PH_3	1200.00	1200.00	1200.00	1200.00	1200.00

3. Result and discussion

3.1. Structure identification

For the optimization process, a complete positive SIMS mass spectrum was recorded on different interfaces of the complete operative structure for to determine both the species in the sample and possible interferences (mass spectra do not showed here). The basic structure of the operative cell was reported previously [2], the diverse substructures have been identified by meaning of the intensity variations of the main elements across the interfaces.

Figure 1 shows the depth profile of the complete and operative structure, with the signal intensity of each element normalized. The different substructures and interfaces between subcells (GaInP/Ga(In)As/Ge) in a operative cell, are marked in the figure with dot lines.

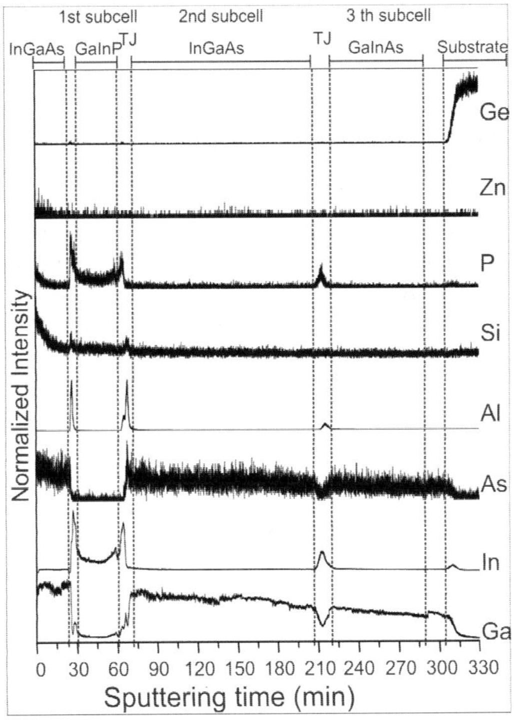

Figure 1. In-depth SIMS analysis of structure similar at operative triple junction solar cell, obtained with O^{2+} ions in positive mode. Subcells and Tunnel Junctions (TJ) are marked with dot lines.

3.2. Study of Diffusion process between interfaces

As mentioned, an operative cell has physical, morphological and optical properties with specific values and must exhibit clear interfaces between the layers of the different structures. In a SIMS depth profile, the interfaces are detected by the variations in signal intensity of the principal elements. Figure 2 exhibits a zoom of the depth profile of the top cell and TJ (Tunnel Junction) zones in samples A and B. Sample A exhibits interfaces between the subcells clearly. The InGaAs/GaAs interface (t= 14.28 min) is identified by the reduction of In and increment in Ga and As. The GaAs/AlInP interface (t=15.30 min) is determined by the meaning the increment of Al, In and P. In general, in sample B is observed a displacement in the sputtering times, as well as blurred interface zones, evidencing some Al diffusion over the InGaAs/GaAs and AlInP/GaInP interfaces. Although the interface GaAs/AlInP (t=13.20 min) in sample B appears at the same sputtering time than in sample A, the Al signal is observed last longer than in the sample A and the structure is observed enlarged.

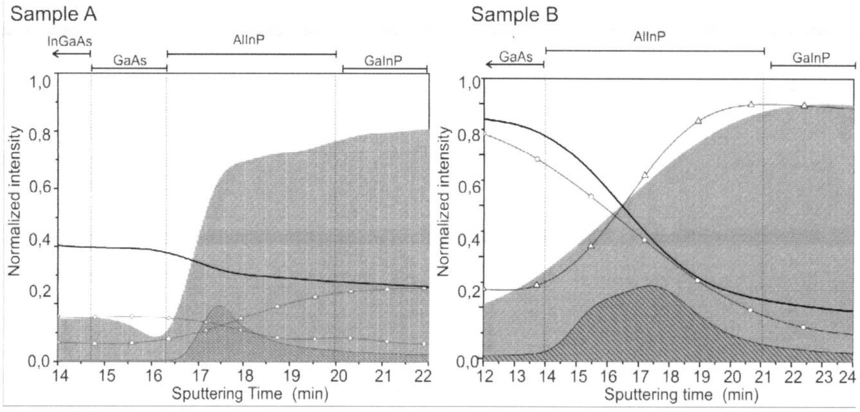

Figure 2. Detail zoom of the in-depth analysis of hyperstructure corresponding of top cell Sample A (left) and Sample B (right): contact layer (GaAs/InGaAs), window (AlInP) and emissor (GaInP). Main elements: Ga (solid line), In (grey), P (triangles), As (circles), Al (dashed) and Ge (squares).

Figure 3 shows the depth profile of interfaces between the top and middle cell in sample A and B. Again, significant differences are observed in the defective cell depth profile. In Sample B profile the AlGaInP and GaAs/AlGaAs layers are significantly thicker than in the operative cell, is indicating by the lack of the bump in the In signal and the much broader diffusion of Al. Diffusion processes are also observed in the interface between top and middle subcell, the interfaces are not clearly defined and the sputtering times are modified. The diffusion of Al and In on next substructure blurring the interfaces AlGaInP/GaAs/AlGaAs, in a similar process observed over the upper section (Figure 2). Similar behaviour is detected in the union between the middle and bottom cell, it is shown in figure 4. In sample A (Figure 4 left), the GaInP/AlGaInP interface is observed at t=130 min, the AlGaInP/GaInAs interface is detected at t=140 min. In the sample B (Figure 4 right) GaInP/AlGaInP interface appears at 128.5 min, the AlGaInP/GaInAs interface is detected at t=150 min; its observed Al diffusion over the bottom cell and the In and As profiles are enlarged and undefined.

The interaction of the sample with the sputtering primary ions can be result in small diffusion in the samples or ion implantation [6]. Diffusion processes are not observed in sample A where the structures are clearly defined while in the sample B it causes enlargement of the structures and degradation of the interfaces. The causes of variation of the sputtering times of sample B, with respect at sample A, are not totally understood. Based on the results, SIMS analysis suggests that the diffusions observed in the sample B are a consequence of the manufacturing process causing defects in the sample. It have been reported the mutual diffusion and atomic exchange of species at the heterointerfaces (InGaP/GaAs) [5], As diffusion over InGaP layers grown with MOVPE [7] and the interdifussion of Al in thin structures [8]. All these processes smooth out the interface abruptness and degrade device performance. We assumed, in agree with the literature and SIMS analysis, that the incorrect incorporation of Al and In at the structure during growth process cause the observed inter-diffusion process and affects at the rest of element profiles, causing degradation and reducing the quality performance of the sample B.

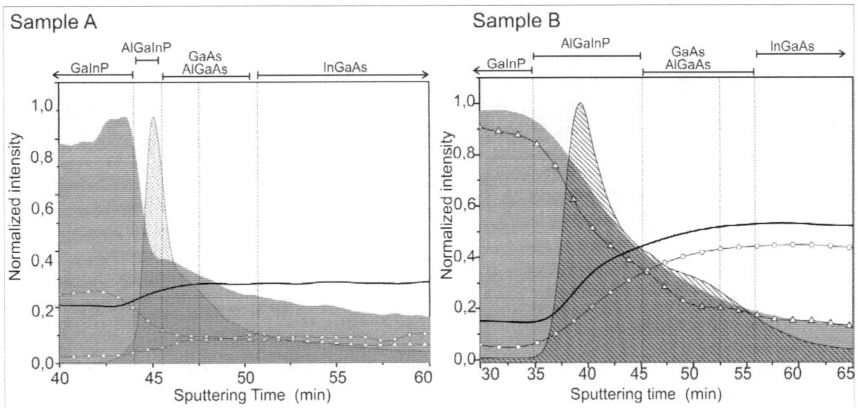

Figure 3. Detail zoom of the in-depth analysis of heteroface between top and middle cell of sample A (left) and sample B (right): Base top cell (GaInP), tunnel junction (AlGaInP/AlGaAs/GaAs/AlGaAs) and emissor middle cell (InGaAs). Main elements: Ga (solid line), In (grey), P (triangles), As (circles), Al (dashed) and Ge (squares).

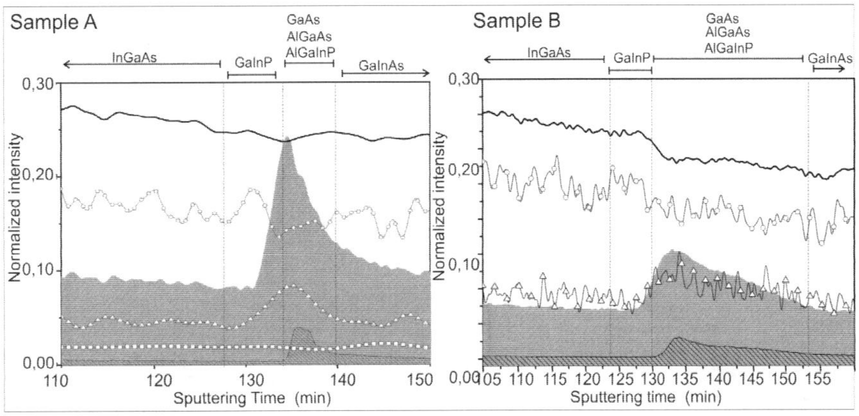

Figure 4. Detail zoom of the in-depth analysis of heteroface between middle adn bottom cell of sample A (left) and sample B cell (right): Base middle cell (InGaAs), tunnel junction (GaInP/AlGaInP/AlGaAs/GaAs/AlGaInP) and botton cell (GaInAs). Main elements: Ga (solid line), In (grey), P (triangles), As (circles), Al (dashed) and Ge (squares).

3.3. Influence of fabrication parameters

SIMS allows study the influence of the changes in the growth conditions in the structure composition and properties. Changes in the reactor flows during the growth process cause variations in the structure, dopants distribution and the cell performance [4]. It was studied a set of simple structures grown over GaAs substrate. SIMS analysis was made in positive mode, with Ar^+ ions and current of 580 nA following the procedure described in [2]. The in-depth profiles of samples are similar for all samples and there are not apparent differences between samples (Figure 5). Ga and As signal are stable at t=14 min, indicating the reaching of the substrate.

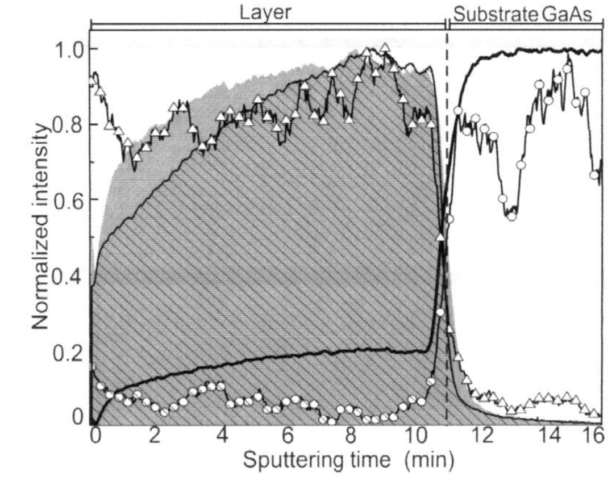

Figure 5. In depth-analysis of basic structure over GaAs substrate. Main elements: Ga (solid line), In (grey), P (triangles), As (circles) and Al (dashed).

Normalization to an internal reference element facilities the direct comparison of different profiles, assuming the same material and properties and allows the determination of differences between the samples [6]. Considering the samples, Al is selected as its growing parameters were not changed in any of the samples, and its SIMS signal is high. The performance of SIMS signal to reveal differences in the growth layers as a consequence of changes in the manufacturing process may be seen in figure 6. As seen in the interface layer/substrate (t=11 min), there are changes in the sputtering time and in the slope of the signal. Sample 2 (with a higher flow of GA during the growing process) shows a higher sputtering time and its slope is sharper than the other samples (Figure 6 right). The variation in the sputtering rate could indicate that the layer physical properties have been modified and that the layer diffusion over substrate is reduced with an increment in the Ga flow.

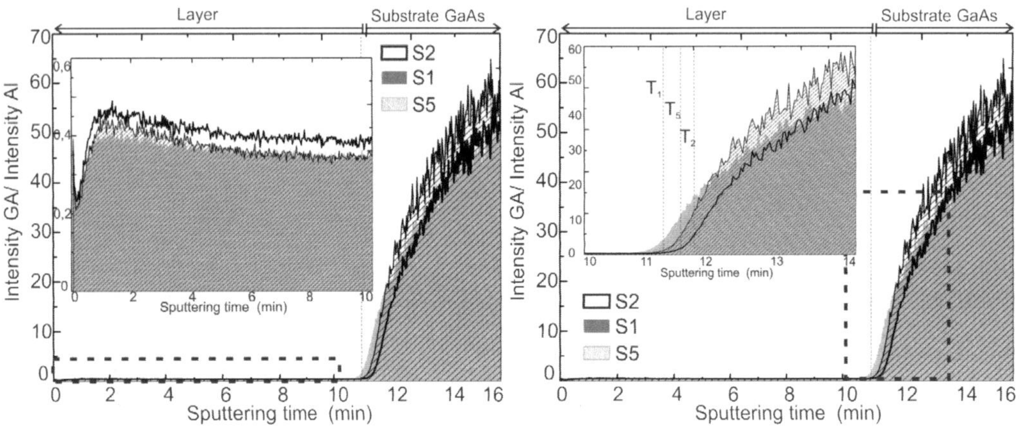

Figure 6. In-depth analysis of Ga signal normalized with respect at Al signal of samples: S1, S2 y S5.Left: Zoom of Layer zone. Right: Zoom of interface zone.

4. Conclusions

SIMS has been used like to study triple junction solar cells and other simple structures grown by MOCVD. With the elemental depth profile (Ga, As, In, Al, P and Ge) allow the clear identification and assignment of the substructures and the functional layers in an operative cell. The depth profile of defective cell shown diffuse interfaces between layers and blurring the structure, it can be associate at inter-diffusion caused by the incorrect incorporation of Al and In at the structure or failures in the process during growth process. The influence of the manufacturing parameters in the layer properties has been studied in simple structures, and it is determined that the changes in the gas flow of organic compounds in the MOCVD reactor can cause variations in the layer diffusion or layer properties. The SIMS capacity for detect diffusion or incorrect incorporation of elements in the complete hyperstructure and detect failures during the manufacturing process has been established. SIMS has

IOP Conf. Series: Materials Science and Engineering **59** (2014) 012002 doi:10.1088/1757-899X/59/1/012002

demonstrated to be an extremely sensitive tool to evaluate the growth process in a different type of hyperstructures.

Acknowledgments
The authors acknowledge the financial support of this work through project CTQ2008-02197 and CTQ2007-60348 (Ministerio de Ciencia e Innovación, Madrid).

References

[1] Luque A, Andreev V, 2007 *Concentrator Photovoltaics.* (Springer, Berlin) p. 1-23
[2] Téllez H, Vadillo J M, Messmer E R, Miguel-Sánchez J and Laserna J J 2011 *Surf. Interface Anal.***43** 646-48.
[3] Galiana B, Volz K, Rey-Stolle I, Stolz W, Algora C 2006 *Proc. of 4th World Conference on Photovoltaic Energy Conversion*
[4] Alferov Z I, Andreev V M, Rumyantsev V D, 2009 Book High-Efficient Low-Cost Photovoltaics **140** (Springer) p 101-41.
[5] Bosi M and Pelosi C 2007 *Prog. Photovolt: Res. Appl.*1551-68
[6] Durose K, Asher S E, Jaegermann W, Levi D, McCandless B E, Metzger W, Moutinho H, Paulson P D, Perkins C L, Sites J R, Teeter G, Terheggen M 2004 *Prog. Photovolt. Res. Appl.* **12** 177
[7] Pelosi C, Attolini G, Frigeri C, Bersani M, Giubertoni D, Vanzetti L and Kudela R 2004 *The Eur. Phys. J. Appl. Phys.* **27** 379-83
[8] Liu W K, Santos M B 1999 *Thin Films: Heteroepitaxial Systems* (World Scientific, Singapore) p 401.

Corrigendum

Study of Metal Organic Chemical Vapour Deposition (MOCVD) semiconductors III-V hyperstructures with Secondary Ion Mass Spectrometry (SIMS).

The auhthor and affiliations were not produced in the correct order in the original article. The correct order is below.

D Padilla-Rueda[1], H Téllez Lozano[2], José M Vadillo[3] and J J Laserna[3]

[1]Physics Departament, Universidad Industrial de Santander, UIS, Cr 27 calle 9 Ciudad Universitaria, Bucaramanga, Colombia
[2]International Institute for Carbon-Neutral Energy Research (WPI-I2CNER), Kyushu University, Japan
[3]Universidad de Málaga, Departamento de Química Analítica, Facultad de Ciencias, 29071, Málaga, España-Spain

IOP Conf. Series: Materials Science and Engineering **59** (2014) 012003 doi:10.1088/1757-899X/59/1/012003

Structure of poly(lactic-acid) PLA nanofibers scaffolds prepared by electrospinning

E Y Gómez-Pachón[1]*, R Vera-Graziano[2], R Montiel Campos[3]

[1] Escuela de Diseño Industrial, Universidad Pedagógica y Tecnológica de Colombia-UPTC, Seccional Duitama, Carrera18 Calle 22, Duitama, Boyacá, Colombia
[2] Instituto de Investigaciones en Materiales, Universidad Nacional Autónoma de México, Avenida Universidad 3000, Delegación Coyoacán, 04510 México, D. F., México[3] Departamento de Polímeros, Universidad Autónoma Metropolitana, México D. F, México, 09340 México DF, México

E-mail: edwin.gomez02@uptc.edu.co, edwinyesidgom@yahoo.com.mx

Abstract. The structural properties of poly(lactic-acid) PLA nanofiber scaffolds prepared by electrospinning have been correlated with their process condition. The influence of the electrospinning processing parameters on structure including fiber orientation, take-up velocity and post-thermal treatment was analyzed. The structure and the properties of the scaffolds were studied by x-ray diffraction (XRD), atomic force microscope (AFM), scanning electron microscopy (SEM), high resolution transmission electron microscopy (HRTEM), and differential scanning calorimetry (DSC). The existence of crystallinity into the nanofibres of PLA was found. The careful observation by HRTEM shows an amorphous core and a semicrystalline shell structure (supramolecular), and the surface of nanofiber observed by AFM shows a laminate periodic along the main axis of the fiber. These observations will be useful in understanding the structure–property relationships of oriented nanofibre scaffolds for medical or biological applications.

1. Introduction

The use of scaffolds and porous materials produced by several techniques are more common nowadays for tissue engineering [1]. A global tendency is the use of biodegradable biomaterials, similar to native tissues, composed by fibers like collagen, which might promote the cell growth and the extracellular matrix generation [2-4]. In order to achieve this goal, several techniques have been used to obtain scaffolds such as drawing, template synthesis, phase separation, self-assembly, biocomponent extrusion and electrospinning among others. Among the several techniques to produce scaffolds, electrospinning is one of the most versatile alternatives for controlling the structural parameters of fibrous scaffolds such as diameter, texture, porosity and fiber orientation [4-7].

[1]* To whom any correspondence should be addressed.

An interesting work about the effect of crystallinity of PLA microfiber by electrical and centrifugal field on mechanical responses was described by Liao et al. [8]. Lim and Tan reported that electrospun polycaprolactone (PCL) nanofibers with smaller diameters had a higher degree of molecular orientation, crystallinity and mechanical properties [9]. In other paper, Tan and Lim reported the effect of annealing on electrospun poly(L-lactic acid) (PLLA) about crystalline structure after electrospinning [10]. They concluded that annealing was found to increase the Young's modulus of the nanofiber due to the increase in crystallinity and the change in morphology. It was clear that the fiber morphology affects the mechanical response of these materials. Wong et al. deduced an enhancement of molecular orientation and an abrupt shift in the elastic modulus of electrospun PCL occurred at around 700 nm in diameter [11].

Naraghi et al. studied the effect of diameter of polyacrylonitrile (PAN) nanofibers about the improvement of molecular orientation and mechanical properties [12]. Their results pointed out that molecular orientation is created at the late stages of the electrospinning process when the polymer jet viscosity increases and bending instabilities take place. Baji et al. also reported the influence of electrospinning parameters on the morphology and microstructure of several nanofibers (PLA, PCL, polyoxymethylene, nylon, cellulose acetate and other materials) and, in consequence, on their mechanical properties [13]. Yoshioka et al. studied the structure of polyethylene (PE) fibers by transmission electron microscopy (TEM) and the fiber structure was related to the fiber diameter. Therefore, they proposed three different structural models 1) the random-oriented crystalline 2) the shish-kebab and 3) the fibrillar structure [14].

Inai et al. have carried out tensile tests of single electrospun PLLA nanofibers collected in a rotating disc at different take-up velocities (tangential velocity produced by the rotating drum or disc) and found an improvement on the mechanical properties. Also, they reported the effects of the conductivity of the solution and the polymer concentration on the structure of PLLA electrospun [15]. Arinstein et al. suggested that nanofibers have a supramolecular structure composed by different phases [16]. Cicero et al [17] studied fibers of PLA produced by two-step melt-spinning. They found that structural properties can be widely manipulated through a combination of draw ratio and temperature. Using AFM, the fiber morphology as found to be highly fibrillar.

In our perspective there are fundamental issues about the internal fiber structure that must be taken into account such as the hierarchical structure. It has been suggested [16-18], but not clearly proven, that electrospun nanofibers can be visualized as composite materials, because they ~~likely~~ present an amorphous core and a supramolecular (semicrystalline) shell.

The internal structure and morphology of nanofibers of PLA has not been determined with enough precision because the nanometer dimensions. This work focuses on the effects of the velocity take-up and annealing on the microstructure of nanofibers of PLA. These experimental studies about microstructure of nanofiber of PLA and scaffolds structures was supported by a plausible inference of the experimental evidence from X-ray diffraction (XRD), differential scanning calorimetry (DSC), atomic force microscopy (AFM), scanning electron microscopy (SEM), high resolution transmission electron microscopy (HRTEM).

2. Materials and methods

Poly(lactid acid) of Mw 230000 (NatureWorks 2002D LLC, MN USA) was dissolved in 2, 2, 2-trifluoroethanol (Sigma Aldrich) to obtain a 16 % w/v of PLA solution, using magnetic stirring. The PLA solution was electrospun to obtain both random and uniaxial oriented nanofiber scaffolds by using two different types of collector systems separately. The electrospinning process was performed at room conditions, at a flow rate of 0.6 ml/h and at an applied voltage of 15kV (Power supply Spellman, USA) and a digitally controlled pump (KD Scientific, USA). The collector for the random scaffolds was a 10 cm square aluminum plate. The collector to obtain oriented nanofibers is an aluminum cylinder custom-designed that rotate at a predetermined velocity in the range of 1 to 10000 RPM, providing tangential velocities (take-up velocities) ranging between 1 to 3142.6 m/min (52.36

m/s). The nanofibers were collected at 1100 m/min and 1217 m/min (18.3 m/s and 20.28 m/s, respectively). Subsequently, the electrospun scaffolds were kept in desiccators at room temperature. Some scaffolds were annealed at 80 °C during 10 hours (Scorpion Scientific, A-50980 oven). The annealing temperature was selected based on the crystallization temperature of as-spun aligned nanofibers (around 70°C). In general, annealing takes place at any temperature between the crystallization and the melting point.

The scaffolds structure was analyzed by X-ray diffraction, XRD, (Siemmens D-500, Cu Kα radiation; λ=1.5406 Å), differential scanning calorimetry, DSC, (2910, TA Instruments), scanning electron microscopy, SEM, (JEOL-JSM-7600F), atomic force microscopy, AFM, (Nanoscope III, Digital Instruments), and high resolution transmission electronic microscopy, HRTEM, (JEOL-2100f). ImageJ software was used to measure the mean diameter of the nanofibres and the orientation of the uniaxially aligned nanofibers (ANs). SEM images (10000 X) at two different locations of each sample were obtained. The diameters of the continuous fibers located at the foreground plane of the micrograph were measured. For the statistical analysis, at least eight points per nanofibre were selected to measure the diameters. The names and abbreviations of the electrospun fibers are shown in the Table 1.

The internal transverse structure of the oriented nanofibres was observed by HRTEM (JEOL-2100f). The samples were embedded in an epoxy resin Epon-812 (Microscopy ScienceR) and cured for 3 days, to impede the fibers from flexing during the cutting process, which was carried out in an Ultracryo Microtome RMCR model MT- 7000. A conventional cutting pyramid was prepared, and fine sections (80-nm-thick) were obtained by means of a diamond blade (DIAMONTR) at room temperature, as the Tg for PLA is 60 °C. To obtain the proper contrast for the HRTEM observations, the samples were placed in a chamber under Ru04 vapour for 4 h.

Table 1. Names and abbreviations of the electrospun fibers

Names	Abbreviation
Uniaxially aligned nanofibers	AN
Uniaxially aligned nanofibers collected at 1100 m/min	$AN_{1100\ m/min}$
Uniaxially aligned nanofibers collected at 1217 m/min	$AN_{1217\ m/min}$
Annealed uniaxially aligned nanofibers	A-AN
Annealed uniaxially aligned nanofibers at 1100 m/min	A- $AN_{1100\ m/min}$
Annealed uniaxially aligned nanofibers at 1217 m/min	A- $AN_{1217\ m/min}$
Randomly oriented nanofibers	RON
Annealed randomly oriented nanofibers	A-RON

The surface of the ANs was observed by AFM (Nanoscope III, Digital Instruments) with a tip model TESP (Si) in the frequency range of 325–382 kHz (Veecoprobes). All images were taken in tapping mode and presented as height and phase images. The samples, 5x5 mm^2, were cut and fixed at the standard AFM sample holder (metal disc). To avoid displacement and mechanical vibration of the nanofibres during the tip movement, a group of a few ANs was approached, and the AFM scan rate was kept below 0.3 Hz.

3. Characterization

The purpose of the characterization was to determine the structure and morphology of the PLA nanofibers and scaffold.

IOP Conf. Series: Materials Science and Engineering **59** (2014) 012003 doi:10.1088/1757-899X/59/1/012003

3.1 External morphology of electrospun nanofibers.

A set of SEM micrographs of PLA electrospun scaffolds is shown in the Figure 1. The degree of orientation of the AN nanofibers was analyzed at two take-up collection velocities; 1100 m/min (AN 1100 m/min) and 1217 m/min (AN 1217 m/min). The Figure 1A and 1B show that almost all nanofibers are aligned along the tangential direction of the cylindrical rotating collector. Both the diameter and degree of alignment of the nanofibers were measured by mean of the software: Image J. The average deviation angle of alignment was around +/- 3 degrees. Therefore, it could be assumed that the nanofibers are uniaxially oriented. The average diameter of AN 1100 m/min nanofibers were 643.8 +/- 50 nm, they showed a high degree of nanofiber orientation (Figure 1A) with smooth and homogeneous surfaces (Figure 1B). Randomly oriented nanofibers RON scaffolds had a greater average diameter: 992 +/- 257 nm (Figure 1C) and the nanofibers showed a rough surface with considerable variation in the diameter along the longitudinal axis (Figure 1D). It can be concluded that the quality of the AN nanofibers is better than RON scaffolds. The rotational collector not only orients the nanofibers but also stretches them to a regular diameter and surface.

The influence of take-up velocity and annealing in the diameter of the nanofibers is depicted in Figure 1E. The RON (zero take-up velocity) presented the highest diameters and standard deviations. The AN $_{1100m/min}$ and AN $_{1217m/min}$ without thermal treatment had diameters of 643 +/- 50 nm and 747 ±117 nm, respectively. These results were compared with those obtained by Inai et al. [15] with AN of PLA collected at 6 m/min and 630 m/min, these had average diameters of 890+/-190 nm and 610+/- 50 nm respectively. It is worth to notice that Inai's experimental electrospinning parameters were different from ours, however, despite of the combination of parameters such as polymer concentration, feed rate, solution electroconductivity, take-up velocity, and type of collectors used in their experiments were different, the AN nanofiber diameters in both works are similar at velocities above 630 m/min. According to our results take-up velocities above 1100 m/min did not influence the nanofiber diameters. In contrast Inai et al. [15], showed that from 6 m/min to 630 m/min (10 fold change in velocity) there is a significant effect in the reduction of the nanofibers diameter.

The annealed AN nanofibers (A-AN) showed the lowest diameter average values. The A-AN $_{1100\ m/min}$ showed an average diameter of 562.8±210, while A-AN $_{1217\ m/min}$ showed an average diameter of 525.8+/- 90.4 nm. By comparing the annealed and non-annealed nanofiber it was found a small decrement in diameter, ranging between 12 and 30% as shown in Figure 1E. This behavior is due the self-assembly of the chain molecules into the nanofiber as a result of annealing. However this effect was not observed in RON.

The figure 2A shows a phase AFM image of AN prepared without annealing at a take-up velocity of 1217 m/min. Two well-defined alternating (periodic) dark and bright regions appeared on the electrospun nanofiber surface. The dark and bright regions correspond to lamellar and amorphous phases, respectively. Two different phases were clearly identified by using the phase AFM image, captured in tapping mode. This could be due to the difference in the sample hardness, density into the aligned nanofiber. Different regions of the sample were analysed by AFM with highly reproducible results. In the annealed samples, these observed features disappeared. Clearly this observation prove the assumption of Ciceron et al. [17] that represented the morphology of PLA microfibers, produced by melt spinning, as a schematic microfibrillar morphology and confirmed that this model can be used in nanofibers. The lamellar fraction $\mathbf{V_2}$ was experimentally measured by AFM and digital image analysis (Image J software), showing an average lamellar volume fraction of 35.7% for AN $_{1217m/min}$.

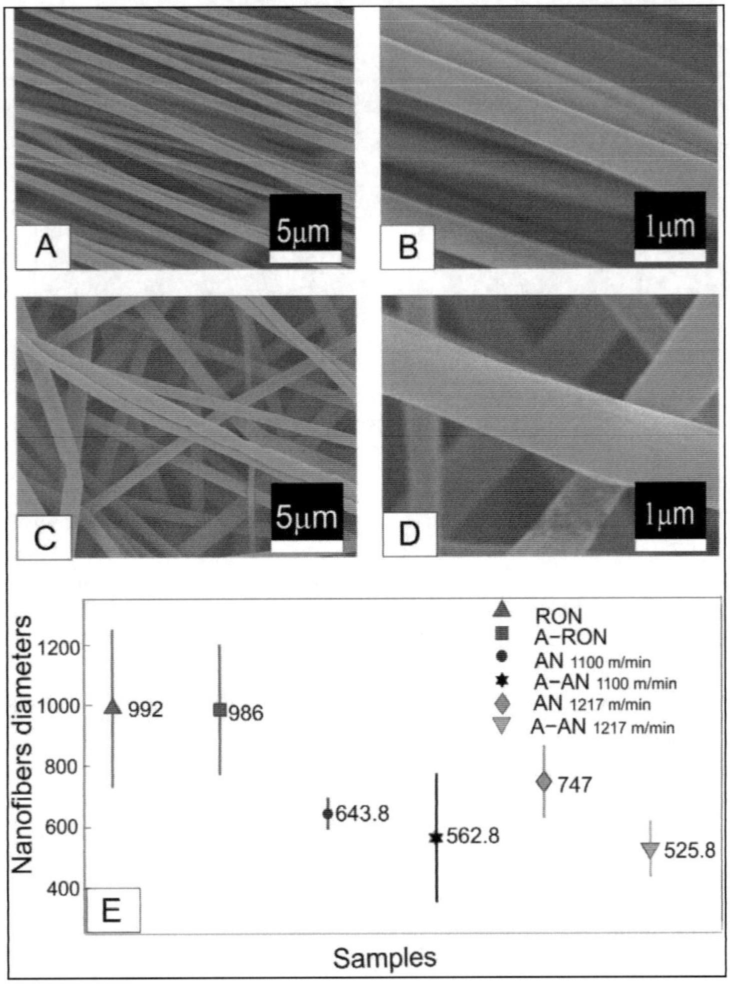

Figure 1. SEM images of electrospun PLA of nanofibers (A) Aligned nanofibers collected at 1100 m/min speed, (B) Same image at higher magnification showing a smooth surface without pores or defects, (C) Random oriented nanofibers collected in a static square plate, (D) Same image at higher magnification showing pores and defects on the surface and (E) Diameters of the different types of PLA nanofibers [random oriented (RON), annealed random oriented (A-RON), (AN) aligned (A-AN) annealed aligned] showing that size was influenced by alignment and heat treatment [18].

The HRTEM image of an AN cross-section, prepared at a take-up velocity of 1217 m/min, is displayed in the figure 2 B, where an amorphous core (bright region) and supramolecular shell (dark regions) phases are clearly distinguished. This confirms the assumption of Arinstein et al. [16] regarding the formation of oriented polymeric chains (supramolecular structure) in the shell due to the electrospinning process. Until this moment, this figure 2B is the first observation of the hierarchical structure of AN that shows the core-shell morphology. As shown in the figure 2A the shell is composed of a supramolecular lamellar phase.

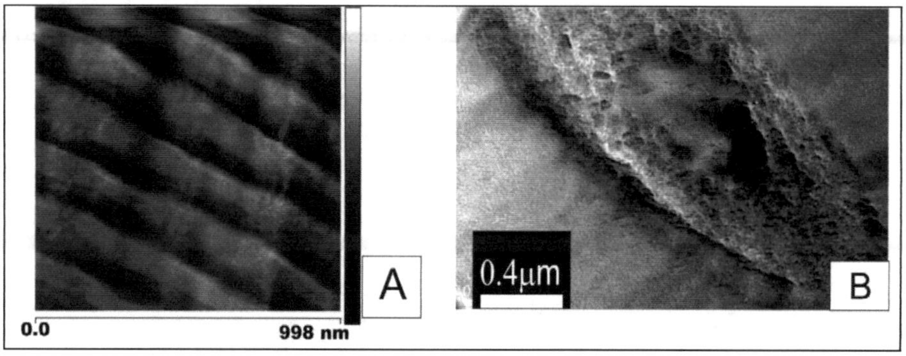

Figure 2. Structure of an aligned PLA electrospun nanofiber (A) Phase AFM image of the nanofiber surface and (B) HRTEM micrograph of the cross section of the nanofiber.

3.2 Crystallographic analysis

The XRD plots (Figure 3) suggest that the level of atomic organization of PLA is strongly dependent on the processing conditions. Bulk PLA, used as a reference, has a semicrystalline structure (purple plot line in Figure 3A) with two characteristic peaks at 16.8° and 19.2° (Bragg angles), in good agreement with those reported by Inai, et al. [15]. These results suggest a crystalline orthorhombic structure (according to PDF-2-2006 card number 00-054-1917, Diffract Plus 2005), with the following lattice parameters a=10.61, b=6.05 and c=28.8 Å. The randomly oriented nanofibers (RON) have a typical amorphous structure (green plot line in Figure 3A). The fast solidification of the fibers during the electrospinning process limited the long-period three-dimensional order at the atomic level of the PLA; therefore, no evidence of extensive crystallization was found. However, after the fibers were annealed below the melting point (A-RON) a peak at 16.8° appears (red plot line in Figure 3A). This result is a clear evidence of the self-assembly of the polymer chains at the atomic level.

Figure 3B shows the XRD results of uniaxially aligned nanofibers (AN). There is no evidence of crystallinity in $AN_{1100\ m/min}$ and $AN_{1217\ m/min}$ as depicted by the blue and black plot lines. However, as it could be expected, annealing promotes extensive crystallization of the polymer chains as shown in the red and green plot lines, where again in both plots the peak is observed at 16.8º. These results are similar to those presented by Inai, et al. [15].

It is most likely that during the electrospinning, the AN polymer chains are oriented along the main axis of the nanofibers due to both the electrical field and winding at high speeds. However, the fast solidification rate tends to inhibit crystallization of the polymer. Annealing promotes chain self-assembly and, as expected, the uniaxially oriented nanofibers (AN) develop more crystallinity than the randomly oriented nanofibers (RON).

3.3 Thermal transitions

The thermal transitions of the electrospun nanofibers were studied by DSC in order to measure crystallization trends. The thermal transitions of the electrospun nanofibers are shown in Figure 3C and 3D. It can be observed that the glass transition (Tg) and melting (Tm) temperatures are about the same in all cases: Tg=60 °C and Tm=147 °C. However, the crystallization temperature (Tc) was affected by the take-up velocity. Figure 3C corresponds to bulk PLA, RON and A-RON samples. In this case the RON and A-RON samples, showed an enthalpic relaxation due to the rapid solidification rate during electrospinning. This enthalpic relaxation temperature occurred at about 65°C (endothermic peak). As expected, the degree of molecular disorder of the chains is higher in RON than in A-RON as shown by the enthalpy values of 6.3 J/g and 5.8 J/g, respectively.

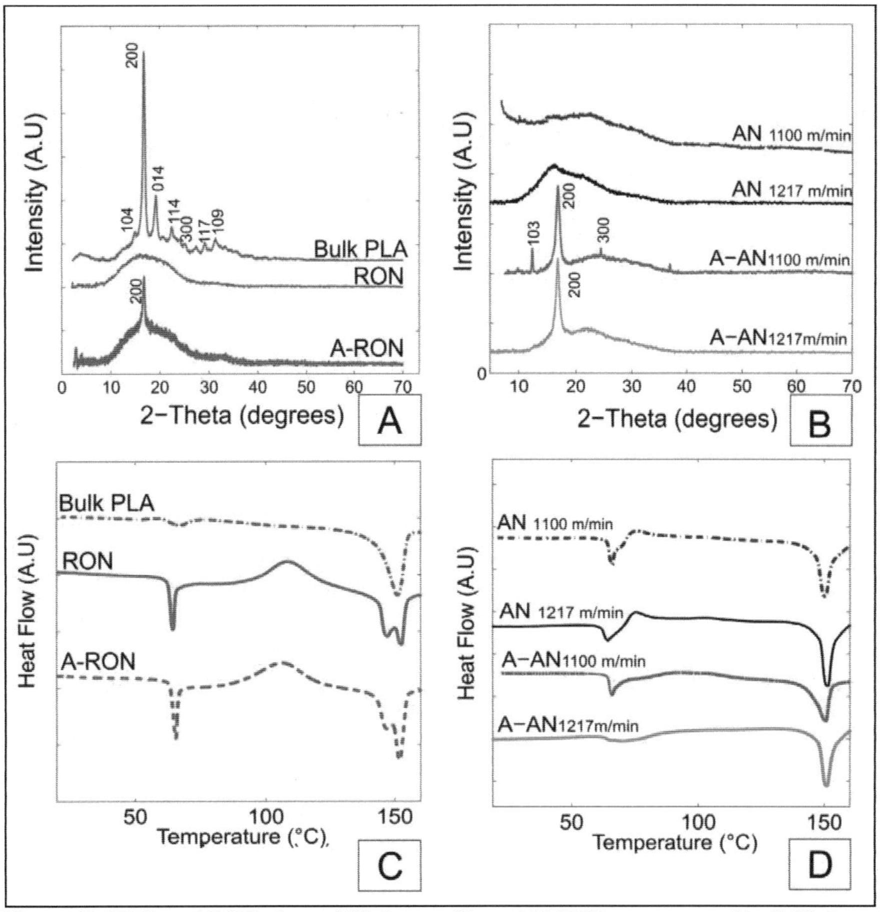

Figure 3. XRD and DSC plots of PLA nanofibers (A) Diffractograms of bulk PLA, RON and A-RON nanofibers, (B) Diffractograms of AN and A-AN nanofibers, (C) DSC of bulk, RON and A-RON nanofibers and D) DSC of AN and A-AN nanofibers [18].

It can be observed in Figure 3C that RON and A-RON plot lines show exothermic peaks between 106-108°C. These peaks, with enthalpy values of 19.4 and 23.1 J/g, respectively, correspond to the crystallization of the polymer. The energy released during crystallization is higher in A-RON because of the effect of annealing at 80°C. Melting is observed at about 147°C. Again the melting peak of A-RON is more intense than in RON with enthalpies values of 25.67 and 21.20 J/g respectively, indicating a greater crystalline fraction in A-RON fibers. The double melting point in RON may be due to two phases coexisting in the nanofibers, because the stretching during the fast electrospinning process may transforms some α crystals into the β form, that possible disrupting the lamellar packing [19].

The thermograms of AN and A-AN also show the enthalpic relaxation peaks again at about 65°C, with enthalpy value of 5.8 J/g for both AN $_{1100\ m/min}$ and AN $_{1217\ m/min}$. This value was similar than that observed in RON. After annealing, the relaxation enthalpies were smaller than in A-RON as follows: 5.8 for A-RON, 4.5 for A-AN $_{1100\ m/min}$ and 3.6 J/g for A-AN$_{1217m/min}$. This tendency was also a confirmation that the combination of take-up velocity and annealing parameters improved the internal organization of the polymer nanofibers. The AN also showed crystallization: in the case of AN $_{1100m/min}$ the crystallization peak was found at 75.3 °C with an enthalpy of 3.9 J/g, while in AN $_{1217\ m/min}$ it was at 74.8 °C with an enthalpy of 4.6 J/g. The orientation of the nanofibers by the effect of the rotational collector induced a significant change in crystallization with respect to RON nanofibers: a reduction of about 30 °C in crystallization temperature (Tc) and smaller crystallization enthalpies, with respect to RON. It implies that AN requires less energy to reach the crystalline state.

The A-AN $_{1100m/min}$ show a Tc= 92.6°C with an enthalpy of 1.5 J/g and A-AN$_{1217\ m/min}$ did not show the crystallization peak. This shows that A-AN reached the maximum value of crystallinity after annealing. The obtained data indicate that annealing has a strong effect on both the crystallization temperature and the crystalline structure of PLA nanofibers. The melting temperatures of AN and A-AN were 146 and 148°C, respectively. The enthalpy values of AN $_{1100\ m/min}$ (28.63J/g) and AN $_{1217\ m/min}$ (31.93 J/g) are greater than those observed in A-RON (25.67 J/g) and RON (21.20 J/g), indicating that AN and A-AN are more crystalline due to the orientation of the polymer chains along the nanofibers.

In the case of A-AN the melting enthalpy was greater for the fibers collected at 1217 m/min than those collected at 1100 m/min. In A-AN $_{1100\ m/min}$ and A-AN $_{1217\ m/min}$ the enthalpies were of 18.47 J/g and 29.84 J/g, respectively. These values indicate that a higher take-up velocity improved crystal formation. Our results showed the crystallization peak, and the XRD analysis confirmed the amorphous-crystalline structure in the samples, as it can be observed in Fig. 3. The electrospinning, nanofiber orientation, take-up velocity, and annealing promote self-assembly of the polymer chains.

According to the evidence of XRD (Figure 3A and 3B), apparently AN nanofibers are not crystalline before annealing. However DSC reveals a crystallization peak in the same type of samples. The XRD technique is not the best to detect the small polymer crystalline structures in the nanofibers in contrast to DSC. In addition to the fact that DSC is a very sensible technique to detect enthalpy changes including crystallization. During the DSC experiments, the heating process induced the oriented polymer chain structures to form bigger crystals before the melting point.

Conclusions

The morphology of scaffolds of electrospun PLA nanofibers depend strongly on the take-up velocity and annealing conditions; winding the electrospun nanofibers in the rotational collector produced highly aligned nanofibers (AN). The morphology of AN was more homogeneous in diameter, without defects, because of nanofiber stretching, than those observed in randomly oriented nanofiber (RON) scaffolds. Fiber stretching produced in turn a high degree of self-assembly.
Molecular orientation increases with take-up velocity; however, the fast solidification occurring during electrospinning limited the size of the crystalline structures. Annealing promoted better molecular self-assembly in the nanofibers, increasing the size of the crystals and the degree of crystallization. Aligned nanofiber scaffolds have a hierarchical core-shell structure where the core is amorphous and the shell has a supramolecular morphology as observed by HRTEM. The surface of nanofibers has two periodic phases: the lamellar and amorphous as observed by AFM.

The morphological and structural features found in the nanofiber scaffolds allow to confirm the structures proposed by other researchers with clear and convincing evidence. Also allow to generate a complete description of the structure of the PLA nanofibers and their relationship with variables such as nanofiber orientation on scaffolding, take-up velocity of the collector and subsequent heat treatments to the spun fibers.

Acknowledgments

The authors thank Nikola Batina and Israel Morales Reyes for AFM analysis, Omar Novelo Peralta for SEM analysis, Patricia Castillo Ocampo for TEM analysis, Adriana Tejeda Cruz for XRD analysis, Damaris Cabrero Palomino for DSC analysis, Alfredo Maciel and Francisco Sanchez for the review of this paper. This project has been supported by ICyTDF (Grants 180/2009 and PICSA12-204), DGAPA-UNAM (Grants IN116809 and IN108913), CONACYT (Grant 129658 and a scholarship), PCeIM-UNAM and DIN-Dirección de Investigaciones-UPTC (Grants SGI 1397/2013) for other supports.

Reference

[1] Barnes CP, Sell S A, Boland E D, Simpson D G, Bowlin G L 2007 *Nanofiber technology: designing the next generation of tissue engineering scaffolds* Advanced Drug Delivery Reviews 59, 1413–1433.

[2] Jayaraman, K., Kotaki, M., Zhang, Y., Mo, X., Ramakrishna, S., 2004. *Recent advances in polymer nanofibers.* Journal of Nanoscience and Nanotechnology 4, pp. 52–65.

[3] Agarwal S and Wendorff J H, Greiner A 2008 *Use of electrospinning technique for biomedical applications.* Polymer 49 pp.5603–5621.

[4] Lannutti J, Reneker D, Ma T, Tomasko D and Farson D 2007 *Electrospinning for tissue engineering scaffolds.* Materials science and engineering c 27 504–509.

[5] Liao S, Murugan R, Chan C K and Ramakrishna S 2008 *Processing nanoengineered scaffolds through electrospinning and mineralization suitable for biomimetic bone tissue engineering.* Journal of the mechanical behavior of biomedical materials 1, pp. 252–260.

[6] McClure M, Simpson DG and Bowlin GL 2012 *Tri-layered vascular grafts composed of polycaprolactone, elastin, collagen, and silk: Optimization of graft properties.* Journal of the Mechanical Behavior of Biomedical Materials 10 pp. 48–61.

[7] Reneker DH and Yarin A L 2008 *Electrospinning jets and polymer nanofibers.* Polymer 49 pp. 2387-425.

[8] Liao C-C, Wang C-C and Chen C-Y 2011 *Stretching-induced crystallinity and orientation of polylactic acid nanofibers with improved mechanical properties using an electrically charged rotating viscoelastic jet.* Polymer 52 pp. 4303-4318.

[9] Lim C T, Tan E P S and Ng S Y 2008 *Effects of crystalline morphology on the tensile properties of electrospun polymer nanofibers.* App. Physics Letters 92 141908-3.

[10] Tan EPS, Lim CT., 2006. *Effects of annealing on the structural and mechanical properties of electrospun polymeric nanofibres.* Nanotechnology 17 pp. 2649–2654.

[11] Wong S-C, Baji A and Leng S 2008 *Effect of fiber diameter on tensile properties of electrospun poly(ε-caprolactone).* Polymer 21 pp. 4713–4722.

[12] Naraghi M, Arshad S N and Chasiotis I 2011 *Molecular orientation and mechanical property size effects in electrospun polyacrylonitrile nanofibers* Polymer 52 pp. 1612-1618.

[13] Baji A, Mai Y, Wong S, Abtahi M and Chen P 2010 *Electrospinning of polymer nanofibers: Effects on oriented morphology, structures and tensile properties* Composites Science Technology 70 703-718.

[14] Yoshioka T, Dersch R, Tsuji M and Schaper A K 2010 *Orientation analysis of individual electrospun PE nanofibers by transmission electron microscopy.* Polymer 51 pp. 2383-2389.

[15] Inai R, Kotaki M and Ramakrishna S 2005 *Structure and properties of electrospun PLLA single nanofibres.* Nanotechnology 16 pp. 208–213.

[16] Arinstein A, Burman M, Gendelman O and Zussman E 2007 *Effect of supramolecular structure on polymer nanofibre elasticity.* Nature nanotechnology 2 pp. 59–62.

[17] Cicero J A and Dorgan J R 2002 *Physical Properties and Fiber Morphology of Poly(lactic acid) obtained from continuous two-Step melt spinning* Journal of Polymers and the Environment 9, pp. 1-17.

[18]Gomez-Pachon E, Sánchez-Arévalo F M, Sabina F J, Maciel-Cerda A, Campos M R, Batina N, Morales-Reyes I and Vera-Graziano R 2013. *Characterisation and modelling of the elastic properties of poly(lactic acid) nanofibre scaffolds.* Journal of Materials Science 48, pp. 8308-8319.

[19] Montes De Oca H and Ward I M. *Structure and Mechanical Properties of Poly(L-lactic acid) Crystals and Fibers.* Published online in Wiley InterScience (www.interscience.wiley.com).

IOP Conf. Series: Materials Science and Engineering **59** (2014) 012004 doi:10.1088/1757-899X/59/1/012004

Wind and Solar Energy Potential Assessment for Development of Renewables Energies Applications in Bucaramanga, Colombia.

G Ordóñez[1]*, G Osma[1], P Vergara[1], J Rey[1]

[1] Gabriel Ordóñez es Profesor Titular Laureado, German Osma es Estudiante de Doctorado, Pedro Vergara es Joven Investigador y Juan Rey es Profesor Auxiliar de la Escuela de Ingenierías Eléctrica, Electrónica y de Telecomunicaciones de la Universidad Industrial de Santander-UIS, Bucaramanga Colombia y pertenecen al Grupo de Investigación en Sistemas de Energía Eléctrica, GISEL.

E-mail: gaby@uis.edu.co, german.osma@gmail.com, pvergarabarrios@gmail.com, juanmrey@uis.edu.co.

Abstract. Currently, the trend of micro-grids and small-scale renewable generation systems implementation in urban environments requires to have historical and detailed information about the energy potential resource in site. In Colombia, this information is limited and do not favor the design of these applications; for this reason, must be made detailed studies of the energy potential in their cities. In this paper is presented the wind and solar energy resource assessment for the city of Bucaramanga, based on the monitoring on four strategic points during the years 2010, 2011 and 2012. According to the analysis, is evidenced a significant solar resource throughout the year ascending on average to 1 734 kWh/m^2, equivalent to 4.8 $kWh/m^2/day$. Also, from a wind statistical study based on the Weibull probability distribution and Wind Power Density (WPD) was established the wind potential as Class 1 according to the scale of the Department of Energy of the United States (DOE), since the average speed is near 1.4 m/s. Due this, it is technically unfeasible the using of micro-turbines in the city, even so their potential for natural ventilation of building was analyzed. Finally, is presented a methodology to analyze solar harvesting by sectors in the city, according to the solar motion and shadowing caused by existing structures.

1. Introduction

The design of renewable energy systems requires energy potential assessment to properly size its elements, model the behavior of power flows and design the energy management systems. While in countries like Colombia have been identified significant energy potentials, there are significant barriers to entry for this type of systems as a weak legal framework, the lack of financial incentives and limited information about the resource available [1].

Due this, it is important that the designs of these systems have as priority the search of the best financial options, for what is necessary to have information sources that describe and characterize renewable energy resources in the country. These information sources allow establishing the feasibility of implementing certain types generation in some areas of the country, thus encouraging the development of projects that implement systems based in renewable energy.

[1] * To whom any correspondence should be addressed.

2. Wind and Solar potential assessment in Colombia

The behavior of solar radiation in the country can be described according to the study region: in the Caribbean area, where it is estimated the most potential, it's perceived a daily average close to 5.5 kWh/m^2, with an annual estimate of 2 190 kWh/m^2; in the Pacific area is estimated the lower potential with 3.0 kWh/m^2 daily average estimated, and an annual estimate of 1 278 kWh/m^2. Colombia is the country in the region with the best solar potential, with a daily average of 4.5 kWh/m^2 in almost all the territory [2].

Regarding the wind resource is identified that the best potential is in the Caribbean area, specifically in La Guajira, with a power density of 700 W/m^2 at 10 m and 1700 W/m^2 at 50 m, being the more attractive region from a technical standpoint [3]. Due to the wide availability of resource in this area, is identified wind power as a renewable energy source with better features to achieve a higher percentage of participation in the national electricity system.

3. Climate and geography of Bucaramanga

Bucaramanga is located at 7° 08' N latitude y 73° 08' W longitude. The climate is classified by IDEAM as follows: at low altitude areas of the city as warm dry, and at high altitude areas as dry temperate. The winds mostly come from the north and northeast, with relatively low average speeds and a maximum value of 2 m/s, especially during the months of January and March; the strongest winds occur between the months of August to October with speeds between 3 m/s and 5 m/s [4].

4. Data source

The solar and wind resource of the city will be evaluated by studying meteorological variables such as wind speed and direction, temperature and solar radiation. For this, will be analyzed hourly frequency data of the years 2010, 2011 and 2012 in four strategic points of the city. This information was provided by the *Corporación de la Defensa para la Meseta de Bucaramanga (CDMB)* and the *Universidad Industrial de Santander*.

The first station is called E1 (*Centro*), and is located at downtown, the second station is called E2 (*Real de Minas*), and is located at west, the third station E3 (*Norte*) is located at north, close to the station E4 (*UIS*), located at the UIS central campus.

5. Wind assessment

The wind resource assessment refers to the way in which the characteristics most predominant of wind is studied for wind power generation [5]. The wind is a random variable, difficult to model and predict, but can be described by a probability distribution function (PDF) based on in-situ measurements [6]. FDP allows to characterize, depending on the reliability of the measurements and the model fit, the wind behavior and the probability of occurrence for each speed at the area where is planned to install a wind turbine.

The wind data analysis is performed by Weibull FDP determinate by Equation (1) , which depends of two parameters; k knows as shape parameter and C as scale parameter.

$$f(v,k,C)=\frac{k}{C}\left(\frac{v}{C}\right)^{k-1} e^{-\left(\frac{v}{C}\right)^k} \tag{1}$$

Theoretically, the power increases with wind speed cube [7]. To study the power available at the site is used the Wind Power Density (WPD). As part of the DOE's Federal Wind Energy Program, the Battelle-Pacific Northwest Laboratory has developed a classification zone with respect to its WPD in W/m^2. This classification includes a numerical ranking from the lowest level (Class 1) to the higher value (Class 7) [8]. Areas classified as Class 4 or higher are considered to be suitable for generating at large-and small-scale. While Class 2 areas are considered marginal for the development of these projects and Class 1 are not appropriate areas for investment.

In general, the wind resource is scarce in Bucaramanga. The average wind speeds recorded at all stations do not exceed 1.5 m/s. According to the analysis of the data, a quarter of the year winds have speeds approaching 0.5 m/s. Although wind speeds with great potential are presented, greater than 3 m/s, these are recorded in less than 15% of the total hours per year, so that the installation of large wind turbines would not be viable, since the resource estimate is not significant. The greatest resource is estimated at station E1, with an average speed greater than 1.4 m/s, and the average annual energy greater than 40.8 kWh/m^2. Furthermore, the station E2, has the lowest wind resource, with an average speed of 1 m/s, where over 30% of hours per year, the wind has a velocity below 0.1 m/s and is estimated an energy annual of 25 kWh/m^2, as is shown in Table 1.

Table 1. Summary of the analysis of the main parameters for the study of wind potential in Bucaramanga, at a height of 10 m.

Station	Year	$\bar{v}$ (m/s)	k (-)	C (m/s)	WPD (W/m^2)	E_{wt} (kWh/m^2)	Class
E1	2010	1.42	1.42	1.57	4.89	41.66	Class 1
	2011	1.39	1.42	1.53	4.56	39.96	Class 1
	2012	1.40	1.43	1.54	4.65	40.88	Class 1
E2	2011	1.02	1.00	1.02	2.91	22.69	Class 1
	2012	1.05	1.05	1.07	2.85	25.03	Class 1
E3	2010	1.34	1.21	1.43	5.26	42.81	Class 1
	2011	1.27	1.23	1.36	4.29	37.32	Class 1
	2012	1.37	1.30	1.49	5.00	43.46	Class 1
E4	2010	1.48	0.99	1.48	5.82	45.96	Class 1

Weibull's FDP for E1 has a shape parameter equal to 1.42 and a scale parameter equal to 1.57 (for 2010), which represents a narrow curve, with a little occurrence probability of higher velocities than most probable velocity (0.67 m/s). This station presents the better estimation of WPD, with values near to 5 W/m^2.

For E2, the WPD allowed to observe a not considerable wind resource at west of the city, with values not higher than 3 W/m^2 at 10 m and 18 W/m^2 at 50 m, classified as Class 1 according to the DOE scale. The winds in this area of the city come mostly in SE and SSE. For E3, the most probability velocity remains 0.5 m/s, with speeds up to 7 m/s only in 2% of annual hours.

Finally, the station located on UIS campus presents Weibull's FDP parameters less than unity, so that frequency curve decays exponentially. Therefore it is possible to conclude that the probability of speeds higher than average speed is reduced significantly. In this area the wind resource does not exceed 11.64 W/m^2 at 50 m, and the estimated annual energy reaches to be higher to 45.96 kWh/m^2.

6. Solar assessment

The solar resource assessment requires the statistical characterization of the incident radiation at specific locations on the Earths's surface. This analysis is the key to the development of projects related to solar energy generation, because it allows sizing the solar potential and model the energy performance of these systems [9].

The solar potential assessment requires historical measurements covering several years, because this allows estimate properly the systems' capacity which impact on the design stage, the initial investment and the final generation cost [10].

There are several ways to assess the energy potential, as satellite images characterization of Earth's surface [9], and direct measurement of variables in the study site [10]. For this case, direct measurement data was analyzed in order to calculate, the average annual energy (E_y), given by Equation (2), where R_i, corresponds to radiation for hour i in W/m^2.

$$E_y = \frac{1}{1000}\sum_i^{8760} R_i \quad \left(kWh / m^2\right) \tag{2}$$

The Peak Sun Hours are a factor that allows measure total daily energy received at a study point with respect to a value of 1000 W/m^2, commonly used to describe nominal characteristics of solar energy equipments. Otherwise described, corresponds to the number of hours of exposure per day, with this level of radiation, required to obtain a total energy equivalent to the certain value [11]. According with the criteria presented in [12], values greater than 3 PSH mean that the site is

IOP Conf. Series: Materials Science and Engineering **59** (2014) 012004 doi:10.1088/1757-899X/59/1/012004

considered a suitable location for installation of solar systems. Calculating PSH is determined by Equation (3), where R_i corresponds to radiation for the hour i in W/m^2.

$$PSH = \frac{1}{1000}\sum_{i}^{24} R_i \quad (h) \tag{3}$$

In general, it is observed that the solar resource available in the city shows outstanding levels and therefore suitable characteristics for the implementation of photovoltaic solar panels and other applications related to solar energy. In the two stations with higher average annual energy values (E1 and E2) these exceed in most cases 1700 kWh/m^2, while in the other stations (E3 and E4) values are near to 1600 kWh/m^2. In total there is an average daily value of 4.8 kWh/m^2, which is equivalent to 4.8 PSH. Taking into account the criteria described above is determined that the city is a suitable location for the installation of solar systems. Values with greater relevance related to solar potential assessment are presented in Table 2. Temperature measurements show an average total daily temperature of 22.55 °C and for sunshine hours of 26 °C, and a maximum average value of 29.08 °C and a minimum of 19.05 °C.

Table 2. Summary of the analysis of the main parameters for the study of solar potential in Bucaramanga.

Station	Year	Average Annual Energy [kWh/m^2]	Average Radiation [kWh/m^2]	Maximum Radiation [kWh/m^2]	Minimum Radiation [kWh/m^2]	Average PSH [h]	Average Temperature [°C]
E1	2009	1722.201	4.991	7.636	1.638	4.991	21.08
	2010	1734.032	5.160	9.438	1.197	5.160	20.82
	2011	2016.71	6.074	11.378	1.631	6.074	20.47
E2	2011	1646.23	4.510	6.741	1.335	4.510	23.34
	2012	1715.142	4.686	6.855	1.719	4.686	23.88
E3	2010	1553.477	4.315	7.184	0.91	4.315	24.84
	2011	1673.670	4.624	7.239	1.283	4.624	24.30
	2012	1698.468	4.691	7.157	1.770	4.691	24.79
E4	2010	1186.616	4.599	6.939	1.631	4.599	22.92

7. Methodology for the analysis of photovoltaic solar use

Taking into account the available solar potential in Bucaramanga, and that the city is a mostly residential area, it is considered relevant to analyze the PV generation *in situ*, specifically for the satisfaction of the household energy consumption. Initially, data are established about the description of urban (step 1), such as: climate, energy demand and urban distribution; subsequently, the basic estimation of the PV system for satisfying the household energy demand is done (step 2); and finally, the shadowing effect is considered on the generation in situ (step 3).

7.1. Generic data - citywide

Bucaramanga has an area of 165 km^2, an average temperature of 26°C and a solar radiation of 4.8 kWh/m^2/day. The sun movement affects the north facade during 5 months per year (from April to August – period A) and the other 7 months (from September to March) it affects the south facade; but, it is characterized for a significant solar height for hours with highest sun radiation. Its population is consisted by 147 200 household, in 1 341 blocks, for that the dwelling average area would be 73 m^2.

7.2. Estimation of installed capacity required by dwelling

The monthly generation (MG) depends of the area (A), the number of PSH, the PV panels efficiency (η_P), and the efficiency of the management and energy conversion step (η_{MEC}) and the and it can be expressed according to the Equation (4) [13].

$$\mathrm{MG} = \mathrm{A} \cdot \mathrm{PSH} \cdot \eta_p \cdot \eta_{\mathrm{MEC}} \cdot 30 \tag{4}$$

With the purpose of determine the size of the PV system for compensating the energy consumption per dwelling, it is required to estimate the area by Equation (5). For that, MG is equaled to monthly electrical energy consumption (MEEC); that is 200 kWh. The value of PSH is 4.8 by data shown in the numeral 6. The values of η_P y η_{MEC} are taken from [13] and represent the technological characteristics of panels, inverters and charger available in the Colombian market. The needed area per dwelling is 12.4 m^2. For an installed capacity density of 150W/m^2, the PV system should be of 2 kW; this is equivalent to the fifth part of the area dwelling.

$$A = \frac{MEEC}{\eta_P \cdot PSH \cdot \eta_{MEC} \cdot 30} = \frac{200\dfrac{kWh}{month}}{14\% \cdot 4.8\dfrac{kWh}{m^2/day} \cdot 80\% \cdot 30\dfrac{day}{month}} = 12.4m^2 \qquad (5)$$

7.3. Shadowing effect

Bucaramanga is characterized by a high growing in horizontal property (apartments); thus, to determine the PV potential is needed a shadowing analysis caused by the buildings. In the Figure 1(a) is shown the shadows zones that a building generates for the periods A (from September to March) and B (from April to August). The zone of the period A is to the south and corresponds to the shadows caused by the building when the sun radiation affects the north facade; manner similar but contrary happens with the period B.

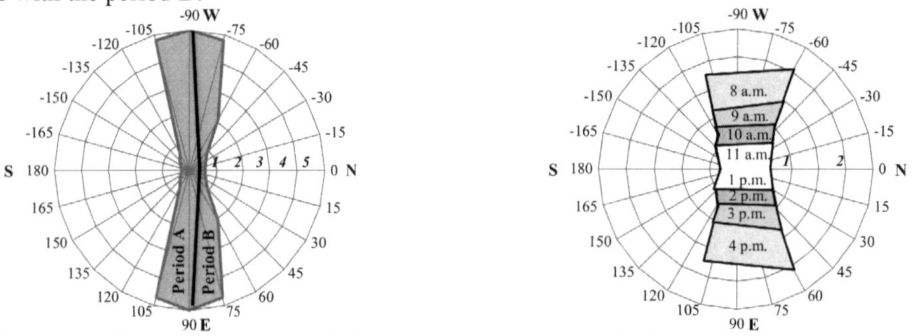

(a). Sunchart of Bucaramanga with Solea-2 (b). Shadowing curves.

Figure 1. Characterization of shadowing curve caused by a building in Bucaramanga.

In the Figure 1(b) these projections on the ground are given in function of the height building with respect to its location. It is established that dwellings on the west and east side of the building only could generate energy of significantly way during half time of solar availability each day, and the zones locates in the north and south side could only generate in normal conditions during half year.

For the case of apartments, the use of PV system can be implemented on roof, but overall in the façades. It is important to clarify that according the sun movement, the level generation in façades would be least to the half in comparison to a system on roof without shadows, so the required area would be larger and surely it is not possible to achieve the self-supply.

8. Conclusions

The wind resource in Bucaramanga has a low energy potential, so it is not recommended the implementation of wind turbines in this area. Although some wind speed may exceed 3 m/s, its frequency is less than 15% of annual hours. Also, it's identified that in over 25% of hours per year the wind speeds does not exceed 0.5 m/s, the measurements carried out on all stations are classified as Class 1 according to the values of WPD.

The solar resource in Bucaramanga has a high potential, so the implementation of photovoltaic panels or systems that take advantage of solar power is convenient. The measurements carried out at all stations exceed the criterion of 3 PSH and is estimated an annual energy close to 1 600 kWh/m^2.

9. Acknowledgement

The authors wish to thank to the *Departamento Administrativo de Ciencia, Tecnología e Innovación* (COLCIENCIAS) and the *Universidad Industrial de Santander (UIS)* for the financial support through the 2013-2013 Young Researchers National Program.

References

[1] D. Ortiz, M. Gualteros, S. Sanabria, and J. Sabogal, "Development and implementation of renewable energies in Colombia as part of the national political agenda: A review of regulations and incentives," in *2012 IEEE PES Innovative Smart Grid Technologies (ISGT)*, 2012, no. c, pp. 1–7.

[2] I. E. Zuniga and Y. F. Botina, "Technical economic study for the implementation of autonomous photovoltaic systems in the residential sector of Colombia," in *2012 Sixth IEEE/PES Transmission and Distribution: Latin America Conference and Exposition (T&D-LA)*, 2012, pp. 1–5.

[3] M. S. J. Joaquin and R. P. Carlos, "Model of application of distributed generation in Colombia rural zones," in *PES T&D 2012*, 2012, pp. 1–9.

[4] R. Lazar and J. Eder, "Estudio sobre el clima urbano en Bucaramanga." Universidad Industrial de Santander, UIS, Bucaramanga, p. 80, 2001.

[5] T. Dai, B. Song, and S.-W. Shu, "Study on the wind energy resources assessment in wind power generation," *2011 2nd Int. Conf. Artif. Intell. Manag. Sci. Electron. Commer.*, pp. 6804–6807, Aug. 2011.

[6] A. Ucar and F. Balo, "Assessment of wind power potential for turbine installation in coastal areas of Turkey," *Renew. Sustain. Energy Rev.*, vol. 14, no. 7, pp. 1901–1912, Sep. 2010.

[7] S. Rehman, S. Shaahid, A. Ahmad, F. Ahmad, T. Thabit, and T. Thabet, "Wind Measurements and Energy Potential for a Remote Village in Saudi Arabia Wind Speed (m / s," no. July, pp. 16–20, 2007.

[8] D. Elliott, C. Holladay, W. Barche, H. Foote, and W. Sandusky, "Wind energy resouerce atlas of the United States." 1987.

[9] O. Gurtuna and A. Prevot, "An overview of solar resource assessment using meteorological satellite data," in *Proceedings of 5th International Conference on Recent Advances in Space Technologies - RAST2011*, 2011, pp. 209–212.

[10] R. A. Suarez, P. Toscano, R. Siri, P. Muse, and G. Abal, "Recent advances in solar resource assessment in Uruguay," in *2012 Sixth IEEE/PES Transmission and Distribution: Latin America Conference and Exposition (T&D-LA)*, 2012, pp. 1–7.

[11] I. Daut, M. Irwanto, Y. M. Irwan, N. Gomesh, and N. S. Ahmad, "Potential of Solar Radiation and Wind Speed for Photovoltaic and Wind Power Hybrid Generation in Perlis , Northern Malaysia," no. June, pp. 6–7, 2011.

[12] S. Ibrahim, I. Daut, Y. M. Irwan, M. Irwanto, N. Gomesh, and Z. Farhana, "Linear Regression Model in Estimating Solar Radiation in Perlis," *Energy Procedia*, vol. 18, pp. 1402–1412, Jan. 2012.

[13] J. M. Rey López and P. P. Vergara Barrios, *Diseño de una microred de baja tensión para el laboratorio de integración energética del Parque Tecnológica de Guatiguará*. Bucaramanga, Colombia: Universidad Industrial de Santander. B. Sc Thesis, 2012, p. 163.

Catalytic effect of Fe, Ni, Co and Mo on the CNTs production.

LM Hoyos-Palacio[1,a], AG García[2], JF Pérez-Robles[3], J González[2], HV Martínez-Tejada[1,b,*]

[1] Universidad Pontificia Bolivariana; Sede Central Medellín; Circular 1 No. 70-01. Medellín, Colombia

[a] Centro de Bioingeniería. Grupo de Investigación en Dinámica Cardiovascular. Facultad de Medicina.

[b] Grupo de Energía & Termodinámica, Grupo de Investigación sobre Nuevos Materiales; Facultad de Ingeniería Mecánica.

[2] Centro de Investigación en Materiales Avanzados, CIMAV. Alianza Norte 202. Parque de Investigación e Innovación Tecnológica. Apodaca, Nuevo León, México. C.P. 66600.

[3] Centro de Investigación y de Estudios Avanzados del IPN. Lib. Norponiente #2000, Fracc. Real de Juriquilla, Querétaro, México.

E-mail: hader.martinez@upb.edu.co

Abstract. In this work a chemical vapour deposition (CVD) process was implemented and automatized for the synthesis of carbon nanotubes (CNTs). Looking the possibility to make an extensive use of biogas, and also for decreasing the costs of the process, methane was used as carbon source. Different CNT structures were obtained using Fe, Ni, Co and also Mo nanoparticles, supported on silica-gel as catalyst for CNT's growing. The number of walls, the morphological effects and efficiency of the process showed dependency on the catalyst type. Cobalt helps the growing of CNTs with fewer layers, compared with Ni, Fe and Mo. Low quality structures are mainly obtained with Fe nanoparticles. Cobalt promotes small diameters in carbon nanotubes and Ni promotes the best quality and yielding during the CVD process.

1. Introduction

Facing the market of nano-materials, one of the main kinds which are increasingly used for different technologies are the CNTs. Due to excellent structural, physical, electrical and mechanical properties, CNTs have stimulated the development of new technologies unimaginable some decades ago. This has become the subject of studies focused on exploring the use of CNTs in various areas [1]. The CNT applications can enable nano-scale molecular levels, which have also generated great enthusiasm in the scientific community and a revolution in modern materials science [2]. Currently is possible to find a variety of studies about CNTs, ranging from structural and physicochemical properties, synthesis to surface modifications [3, 4]. Despite this vast literature, the knowledge obtained on the CNTs are just the prelude of its potential.

Historically, CNTs have been obtained by different methods; including: chemical vapor deposition (CVD), solvothermal (several synthetic strategies have been employed without using transition metal catalyst. Templates, precursors and hydro or solvothermal methods, using binary and complex metal

[1] *To whom any correspondence should be addressed

oxides, chalcogenides, and carbides can be implemented), electrochemical (deposition from organic solutions at room temperature, with formation and growth stimulated by transition metal catalyst, such as iron and/or nickel), or hydrothermal methods (hydrothermal synthesis using polyethylene/water mixtures in the presence of nickel at 700-800° C under 60-100 MPa pressure), as well as arc discharge (soot formed from graphite electrodes during arc discharge), reactive milling (ball-milling and annealing or a pressure-vapor-condenser), laser ablation (a pulsed-laser vaporizes a graphite target in a high-temperature reactor as an inert gas is blend into the chamber), plasma forming (plasma growth at low temperatures) and HiPCO (continuous flow high-pressure carbon monoxide) [5-7].

Although all the methods for nanotube synthesis and processing are still under continuous development, the most common synthesis method is the hydrocarbon pyrolysis process, also known as carbon vapor deposition (CVD). For CNTs production, it is usually known as thermal CVD or catalytic CVD, to distinguish it from many other kinds of CVD used for various purposes [8]. CVD is an economically and industrially viable process due to the facility to scale. In this process, CNTs can grow from catalyst particles using batch or continuous processes such as a fluidized bed. Thermal decomposition of a hydrocarbon vapor is achieved in the presence of the metal catalyst. As a carbon source it is possible to use different hydrocarbons compounds, including biogas, allowing a proper kinetic for the production of CNTs [9] and the modulation of the nanotube diameter [10-11].

The catalysts used for the production of CNTs are usually transition metals supported on silica [12], mesoporous silica (MCM-41, MCM 48, SBA-15), [13], calcium carbonate, zeolites [14] or magnesium oxide [15]. All types of catalysts have been studied with the aim of increasing the productivity of CNTs, for improving their properties and modulate their structures. In this work a CVD process was used for the synthesis of CNTs, using different silica-supported catalysts, like Fe, Ni, Co and Mo. The sol-gel method was used for the synthesis of supported metal catalysts.

2. Experimental procedure

2.1. Catalyst preparation

The sol-gel method has been extensively used in the preparation of supported metal catalysts because it typically results in highly homogeneous materials with high degree of metal dispersion [16]. In this sense, catalysts were supported on silica sol-gel with the metal to 50 percent based on silica added. To obtain metallic catalyst supported on high-surface area silica by the sol-gel method, the polymerization of an alkoxy-silane such as tetrathoxysilane (TEOS), also known as tetraethyl orthosilicate, is carried out in the presence of the appropriate metal precursors. In our case, catalyst nanoparticles were prepared from high purity salts of the transition metals: $Ni(NO_3)_2$•$6H_2O$, $Fe(NO_3)_3.9H_2O$, $Co(NO_3)_2.6H_2O$ and $(NH_4)_6Mo7O_4$. $4H_2O$, from Baker Co [17]. To accelerate the polymerization, an increase in pH can be brought about by addition of a base, which causes a rapid hydrolysis followed by polymerization [18]. Simultaneously with this polymerization process, the metallic ions (Fe, Ni, Co and Mo) precipitate, thus forming a homogeneous and well-dispersed mixture.

2.2. CNTs synthesis

After placing the catalyst inside a quartz tube, a continuous nitrogen flow rate of 1 L/min was passed through the reactor for removing the oxygen. Subsequently, the reduction process was accomplished within at 600° C. The reduction process was kept for 30 minutes in an atmosphere of 90 % v/v of N_2 and 10 % v/v of H_2. Next, the temperature was increased up to 700° C for the nucleation and growing of CNTs.

2.3. Characterization

The high-resolution images were obtained using a high-resolution transmission electron microscope JEOL JEM-2010, operated at 200 kV and a scanning electron microscopy (SEM) JEOL JSM 5300 operated at 5 kV. Complementary RAMAN spectroscopy was performed, and also thermogravimetric analyses (TGA) for CNTs were carried out in a SW Stare Mettler 8.0 device, with a heating rate of 10 °C/min (25 to 900 °C) in an air atmosphere with flow of 50 ml/min.

3. Results and discussion

Raman spectra of CNT samples using different catalysts are presented in Figure 1. Studies carried out by Dresselhaus about Raman spectra for CNT show that the number of defects in graphene layers depends on the ratio I_D/I_G, which must be less than 0.45 for high structure quality of the CNTs [19]. Figure 1a shows the Raman spectra for CNTs-Ni, in which the ratio I_D/I_G is 0.416, relating a high purity material. The highest purity was obtained in CNTs-Co, with an I_D/I_G ratio of 0.26. On the other hand, with Fe, and Mo the quality is decreased in a high level (Figures 1c, 1d), mainly with Fe ($I_D/I_G \sim$ 1.48). This is due to the low solubility of C in Fe. In order to obtain a better quality, in this case the CVD process must performed to high temperatures (~900°C). In our experiments, for comparison purposes, the temperature was always the same for the different metal-catalyst (~700°C).

According to previous reports [20], the increase of the D band intensity (characteristic peak at ~1350 cm^{-1}) with decreasing multiwalled carbon nanotubes (MWCNT) content, is a direct result of the addition of carbonaceous by-products. In the same sense, a decrease in the G' band intensity (characteristic peak at ~2700 cm^{-1}) is observed as the MWCNT mass fraction decreases. The G' band on Figures 1a and 1b reflects the well-structured carbon walls in the samples with Co and Ni catalyst, while the Figures 1c and 1d (CNTs-Fe and CNTs-Mo), indicate a less ordered structures, due to the carbonaceous by-products.

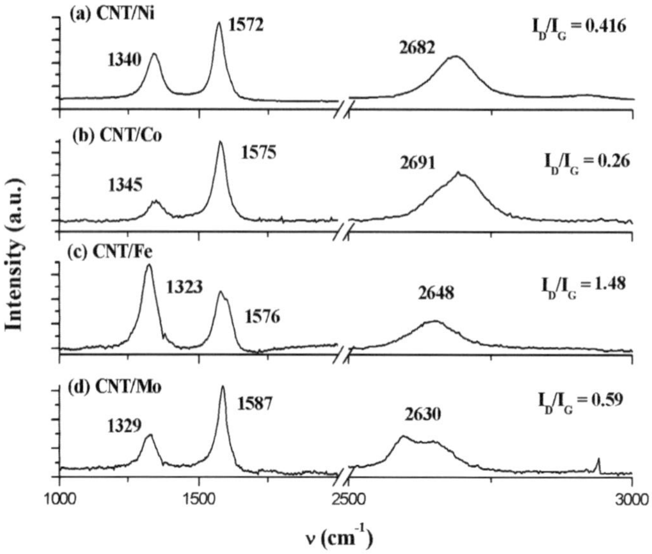

Figure 1. Raman spectra of CNTs grown with (a) Ni, (b) Co, (c) Fe and (d) Mo on silica gel.

FESEM in Figure 2 shows a linear growth of CNTs using Ni as catalyst. The Figure 2a shows that the external diameter distribution of CNTs-Ni ranges from 40 nm to 100 nm. The lengths reached up to several microns (1-7 μm). The tips of these CNTs are closed and contain the catalytic particles enclosed. Figures 2b and 2c show the layers for an isolated CNT. An analysis of the image obtained by HRTEM shown in Figure 2b, lead to conclude that a total of 69 layers and an external diameter of 85.05 nm was obtained (between the layers a gap of ~0,35nm can be calculated in order to obtain the total of layers, according to the methodology developed by Gommes [21] and collaborators, and performing some numerical treatments, through the cross section of the CNTs using the Lambert's law [22]). Also, it is observed that the CNT´s layers are relatively aligned, verifying the structural quality predicted by the I_D/I_G ratio in Raman spectra (Figure 1).

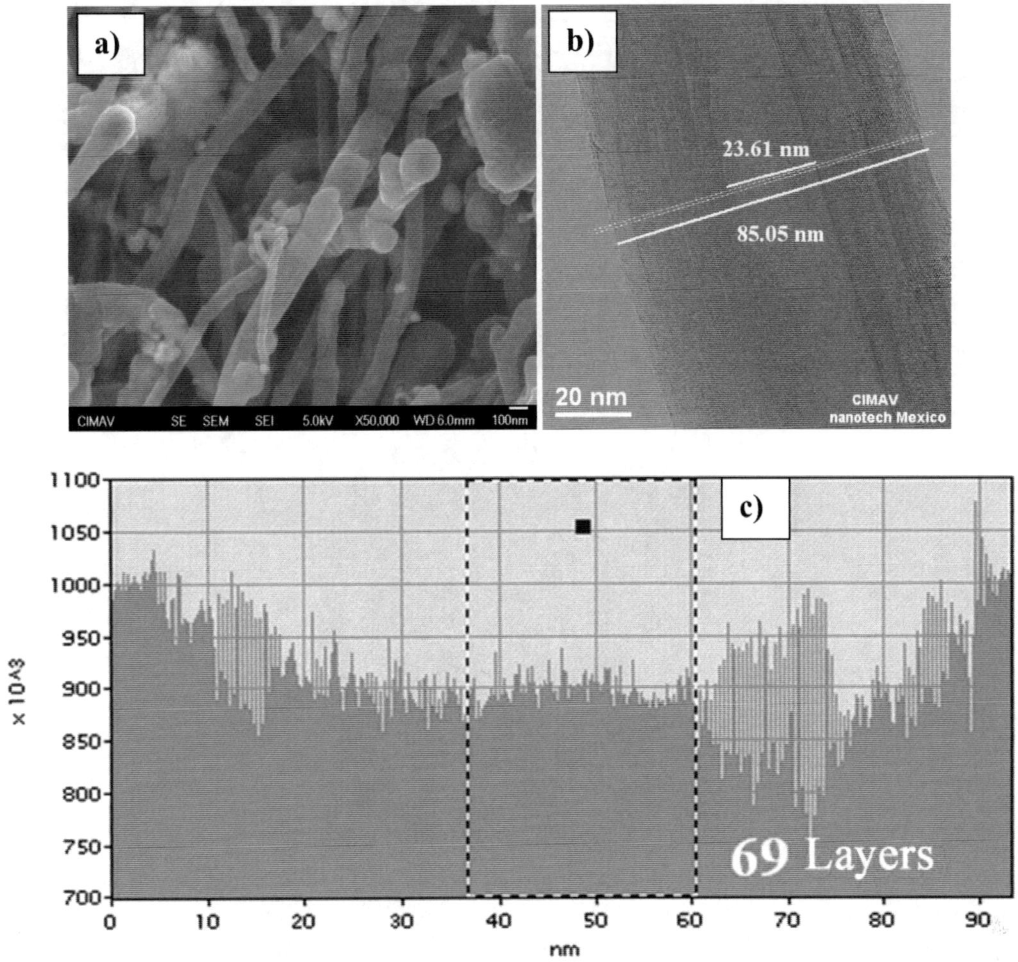

Figure 2. FESEM (a) and HRTEM (b) images for CNT-Ni. (a) and (b) CNT-Ni and (c) number of layers.

Figures 3a-3c show CNT-Co. In this case, the number of walls in the samples is less than the rest of catalyst (Fe, Ni, Mo), with a length above 10 μm. Figures 3b and 3c show that the CNTs layers ranged from 4 to 12. Despite the little number of walls, during the CVD process we obtained MWCNT because the number of walls is in average higher than 4 layers. The CNT-Co allows low number of layers as has been reported by S. Porro [23]. Some CNTs have internal diameters less than 10nm and an external diameter less than 50 nm.

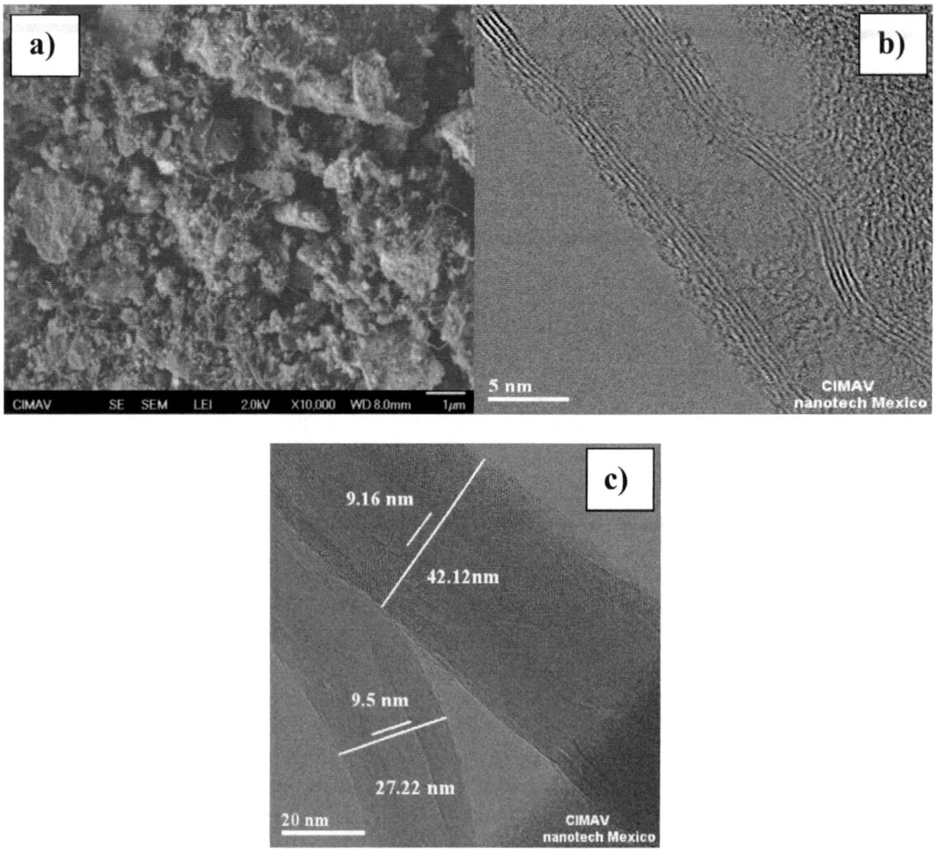

Figure 3. FESEM (a) and HRTEM (b), (c) images for CNT-Co.

Figure 4 shows CNT-Fe with lengths about 4 μm with not high crystallinity, an internal and external diameters of ~ 12 nm and more than ~ 40 nm respectively and 57 layers. The growth of CNT-Fe is tangled and disorderly. According to the results obtained by Huang [24], CNT-Fe and CNT-Co present a non-linear growth. It is also common find the catalyst encapsulated inside the CNT, as a result of the growing process and the little chemical interaction between the catalyst and the silica-gel support. In this sense, the equilibrium state of iron oxide phases (α, β) promote the agglomeration of the catalyst nanoparticles and decrease the encapsulation and the introduction of nanoparticles inside the silica support [25].

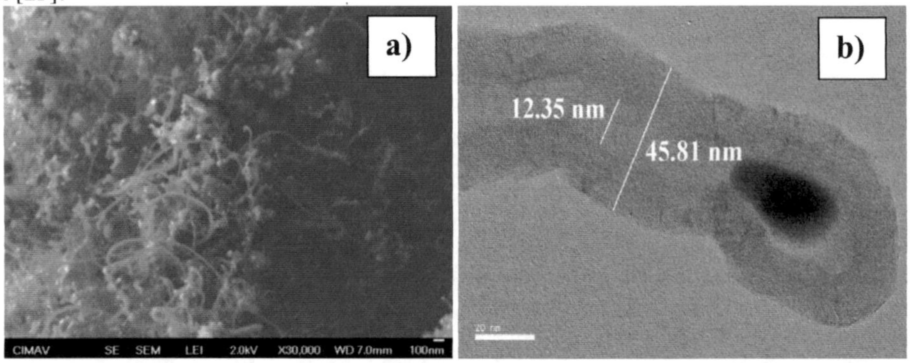

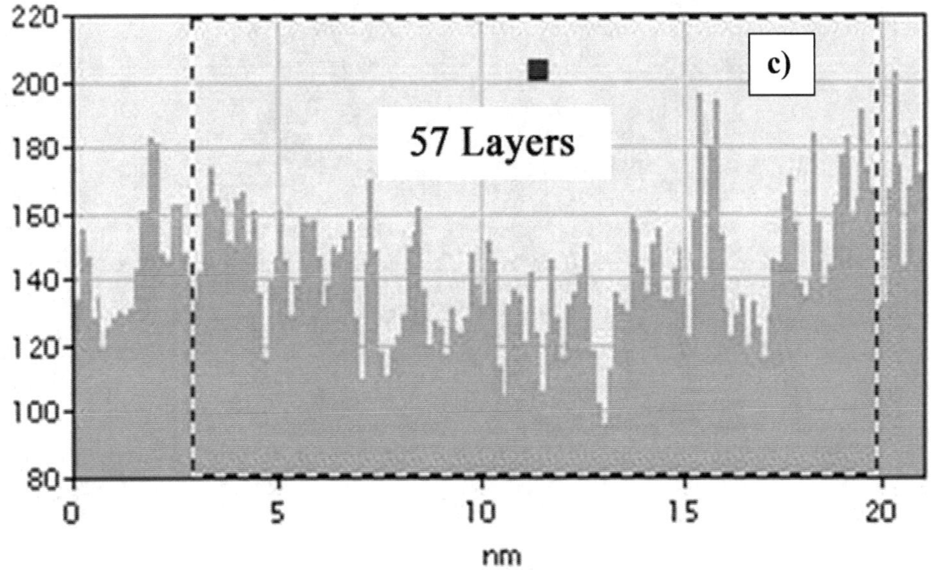

Figure 4 (a) FESEM image of CNTs-Fe, (b) HTEM of CNT. An iron particle encapsulated in a CNT is observed, (c) Number of layers.

Figure 5 shows the CNT-Mo. The micrographs exhibit the low productivity compared with Ni, Co, and Fe. Besides the matrix of silica gel may cover the metal particles, exposing a little surface area to the carbon precursor gas. The main reason for this scarce production of CNTs is attributed to the little solid-state diffusion of carbon in molybdenum. It was difficult to obtain a high purity in samples, dude to the big size of silica clusters (red circle in Figure 5a). The presence of silica during HRTEM analyses inhibits the observation of the numbers of layers due to the electron´s saturation at high voltages (~ 200 kV). Nevertheless, Figure 6 shows the external diameter of CNT-Mo but obtained by FESEM at low voltages. The external diameter ranges between 64.8 to more than 70 nm.

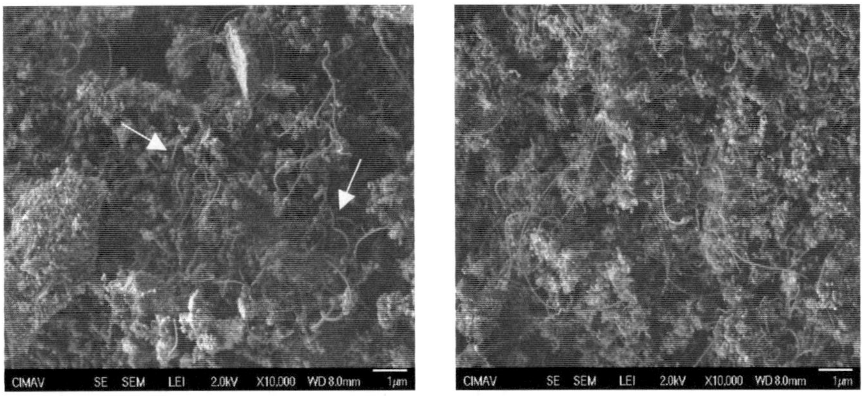

Figure 5 FESEM micrographs of CNT-Mo

Figure 6. FESEM image showing the CNT-Mo center remarked average external diameter range around 64.8 nm.

J.H. Lehman et-al mentioned that no single metric defines nanotube purity. The naïve definition of pure MWCNTs in bulk refers to multiple MWCNTs, all identical, each consisting of a specified number of carbon atoms [26], but this not real during the production of MWCNTs. In this sense, purity refers to the nature of the MWCNT content. On the other hand, for industrial-scale measurements and batch-to-batch quality control, TGA is arguably the most common and practical tool. TGA analysis for CNT-Co in Figure 7a shows an increase in weight since 200 to 400°C, due the oxidation of catalytic metal nanoparticles. Then, the decomposition of carbon structures, related to the loss of weight, is observed from 480 – 500°C. Because some catalytic nanoparticles, encapsulated inside the tubes, can be released during the carbon's structures oxidation; a new increase in weight follows until ~600°C. From this point, the decomposition of different carbon structures continues. In other words, the broad temperature gap (~400°C – ~850°C) indicates diverse number and quality for the carbon layers of nanotubes.

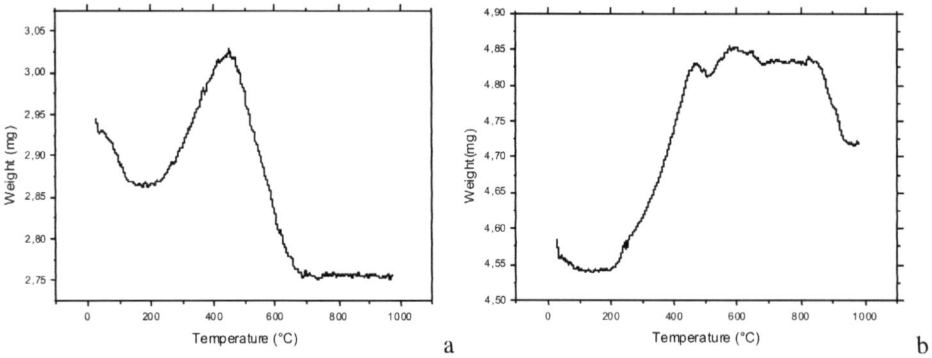

Figure 7. TGA obtained in oxidizing atmosphere for CNT-Co (a) and CNT-Ni (b).

Figure 7b correspond to thermal decomposition of CNTs-Ni samples. In this case, the catalytic nanoparticles were oxidized from ~250 °C, until ~480 °C. After the decomposition of Ni nanoparticles, the oxidation of tubes take place until ~ 700 °C. The linear decomposition observed in CNT samples indicated the high uniformity of the graphene layers between tubes.

IOP Conf. Series: Materials Science and Engineering **59** (2014) 012005 doi:10.1088/1757-899X/59/1/012005

4. Conclusion

When the sol-gel method is used for the synthesis of supported metal catalysts, additional parameters related to the chemistry of the metal precursors are also important. In this regard, this work showed that the different metals used as catalysts allowed the growing of carbon nanotubes. The CNT-Ni is more homogeneous and stable than CNTs grown with other catalysts. Cobalt allows modular small diameter sizes (7-40 nm) during the sol-gel synthesis. The CNT-Fe was not very stable. In other words, because the diffusion of carbon in iron is favored with increasing temperature (>700° C), the presence of amorphous carbon is promoted. The Mo catalyst requires temperatures above 800° C for the deposition of CNTs. The silica sol-gel is a catalytic support that allows high surface area for different catalytic metals (i.e. Fe, Ni, Co, Mo) facilitating the synthesis of CNTs at low temperatures, with different morphological characteristics, for different potential applications.

Acknowledgments
Authors wish to acknowledge the assistance and encouragement from W. Antúnez-Flores, E. Ornelas, and C. Gutierrez at CIMAV Nanotechnology Laboratory in Chihuahua. Also, the financial support from Bilateral Mobility Project CONACYT – COLCIENCIAS (CINVESTAV-UPB), 2009 – 2010.

5. References

[1] Paradise M 2007 Materials and Design **28** 1477
[2] Raffaellea P, Landia B J, Harrisb J D, Baileyb S G., Heppb A F 2004 Materials Science and Engineering **vol** 233
[3] Grobert N 2007 Materials Today **10** 28
[4] Kroto H, W Heath, O'Brien J R, Curl S, Smalley 1985 Nature **138** 162
[5] Terrones M 2004 Material Review **49** 325
[6] Jayasinghe C, Li W, Song Y, Abot J 2010 MRS **35** 682
[7] Terrones M 2003 Annual Review of Materials Research **33** 419
[8] Kumar M, and Ando Y Journal of Nanoscience and Nanotechnology **10** 3739
[9] Raji K., Thomas S, Sobhan CB 2011 Applied Surface Science **257**, 10562
[10] Bo Z, C L, Gang G, Makarovski A, J. L, Gleb F 2002." Nano Letters **2** 895
[11] Jipeng Cheng, Zhiqiang Luo X Z, Fu L, Ying Ye W Y, Wei L, Yuexin H 2006 Materials Chemistry and Physics **95** 5
[12] Huang ZP, D Z W, Wen J.G, Sennett M, Gibson H, Ren Z.F 2002 Appl. Phys. A **74** 387
[13] Urbana M, D M Z, Konya J, Zhub I, Kiricsi 2004 Diamond and Related Materials **13** 1322
[14] Yu a GJ, Gong a J L, Zhu a D Z, He a SX, Cao a JQ, Zhu ZY 2006 Diamond & Related Materials **15** 1261
[15] Coquay P, E D G, Peigney A, Robert E, Vandenberghe, Laurent C 2002 J. Phys. Chem. B **106** 13186
[16] Klein L C 1985 Annual Review of Materials Science **15**
[17] Brinker C J, G W S 1990 Academic Press Inc. pag 185 3d edition
[18] Irurzun V M, Tan Y, Resasco D E 2009 Chemistry of Materials **21** 2238
[19] Dresselhaus M S, G D, Saito R, Jorio A 2005 Physics Reports **409** 47
[20] DiLeo R, Landi A, B J, Raffaelle R. P 2007 Journal of Applied Physics, **101** 1636
[21] Gommes C, Blacher S, et al 2003 Belgum: Carbon **41** 2561
[22] Williams D B and Carter C B 2009 Textbook for Materials Science, Springer 2nd edition
[23] Porro S, S M., Giorcelli M, Chiodoni A, Tagliaferro A 2007 Physica E **37** 58
[24] Hang Qi, D Y, Jie L 2007 The Journal of physical chemistry C letters **111** 6158
[25] Babu P R, W J B, Tyagi P K, Misra D S, Ali N, Gracio J, Cabral G, Titus E 2006 Thin Solid Films **494** 128
[26] Lehman J, H et al. 2011 Carbon **49** 2581.

IOP Conf. Series: Materials Science and Engineering **59** (2014) 012006 doi:10.1088/1757-899X/59/1/012006

Design of energy generation system by using spinning bikes

J Flórez[1]*, C Núñes[1], G Osma[1] and G Ordóñez[1]

[1]Escuela de Ingenierías Eléctrica, Electrónica y de Telecomunicaciones, Universidad Industrial de Santander, Bucaramanga, Colombia

E-mail: florezjulian@correo.uis.edu.co

Abstract. This paper presents the design of a system that generates energy by using the work done by a crew on a spinning bike. The builded system is configured by two subsystems: connection and injection of real-time power to the grid connection and power isolation for specific loads. The first subsystem will have twelve (12) units and the second will have six (6) units.

1. Introduction

Escuela de Ingenierías Eléctrica, Electrónica y de Telecomunicaciones (E3T) from Universidad Industrial de Santander (Bucaramanga, Colombia), has been working on the design of different applications [1] in sustainable energy infrastructure, keeping as goal the reduction of electric power consumption from the public. Having this as main aim, it has come up with a design based on the purpose of contributing to encourage the college community to take care of the environment. This paper presents the design of a system that takes advantage of the kinetic energy developed by the legs, part of the human body with more strength, of a crewman on a spinning bike. Generally, this energy is transformed and wasted as heat. At the beginning of the paper, it will be described an overview of the generation and system sizing. Later, it will be presented an energy analysis applied to the design and the system will be monitored in order to understand its function variables. Then, any financial considerations are presented; and finally it will be shown results and conclusions.

2. Overview about the generation system

Doing exercise by using bikes allows improving metabolic functions, cardiovascular system and respiratory system, it also burns a considerable amount of calories while the user is increasing his muscle tone and mass; in addition, cycling adds an emotional benefit by reducing stress [2]. The mechanical operation of the bicycle requires a boost from a pair of pedals with the legs. Its flywheel is the part of contact in charge of transmitting the energy from the bicycle to the generator. According to [3], the average power generated by a person on a bicycle is between 270 W and 400 W [4].

The use of energy will be given by changing the kinetic energy into electrical energy in the DC generator according to its wide range of operation speed, as the limitations of the commercial supply of generators for these applications [5]. Its limitation will lead on the support structure, called the dock.

It should be taken into account the speed developed by a user, in order to determine the behavior of the electrical output variable. For doing this it should be known: (1) the power curve - speed characteristic of the generator and (2) the relationship between the generator shaft speeds and the

[1] *To whom any correspondence should be addressed.

wheel speed. The PV curve is obtained from the data sheet generator, while the relationship between the speeds requires knowledge of the diameters of the involved parties, also the behavior of the user's pedal speed, because it will be conducted a field test to determine its behavior.

2.1. Performance of a user type

The field test was conducted at the Fitness Center Bodytech in Bucaramanga, during an Indoor Cycling [6] class, which lasted 50 minutes. It was taken by three people in average physical condition. Figure 1 shows the cumulative histogram interval and the wheel speed of the bicycle. It shows a dominant range of speeds, 200 rpm to 250 rpm. The average speed is close to 242 rpm, and it will be above 200 rpm for 80% of the time.

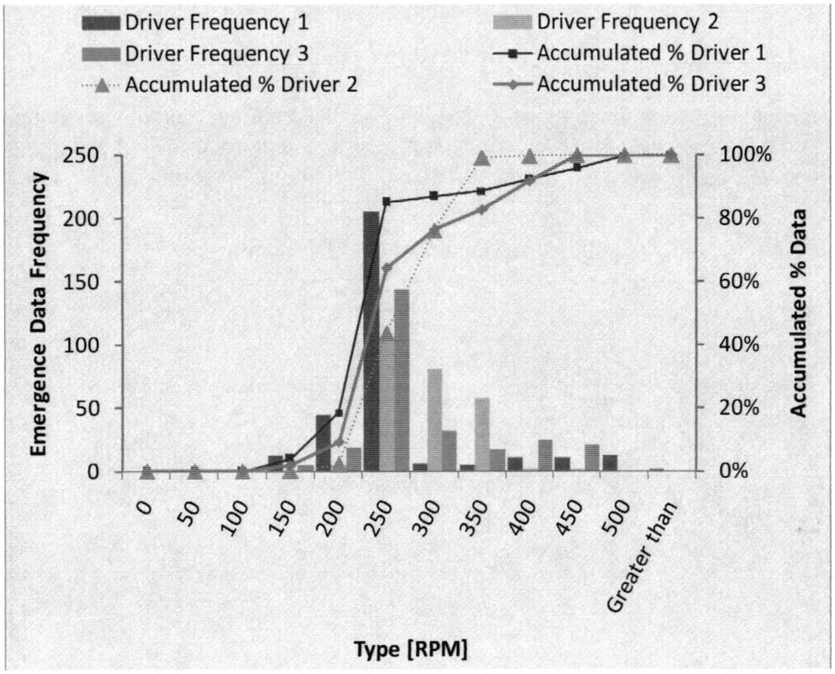

Figure 1. Histogram of the wheel speed of the bike during a spinning session.

2.2. Selection of spinning bicycle and coupling-generator

The bike was a Spinning Stainless previously selected based on the price and technical characteristics from AHP (Analytic Hierarchy Process) [7]. The generator shaft speed and steering wheel of the bicycle are related to its diameters, with a ratio 1:12. Therefore, it is the dominant operating range 2 400 rpm to 3 000 rpm and an average of 2 840 rpm, for which electrical power is close generated to 300 W, as in Figure 2.

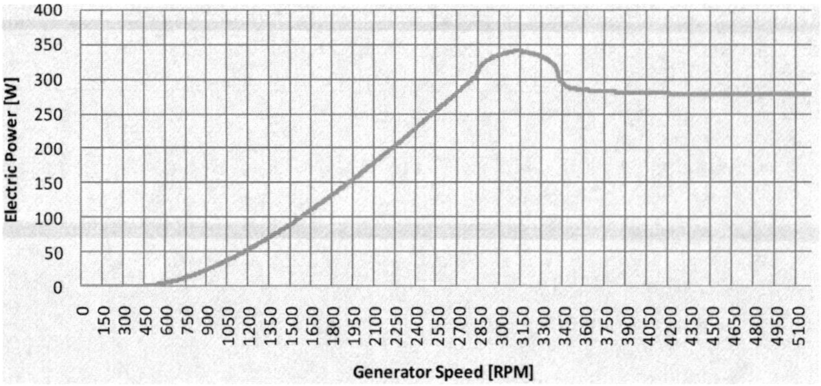

Figure 2. Maximum power vs. RPM (direct drive generator and the crew) [8].

The power generated for a given speed will depend on the strain imposed by the Management Unit and Power Conditioning (Grid Tie Inverter for the subsystem connected to grid and regulator for the isolated subsystem). Table 1 presents the characteristics of a "direct" coupling-type.

Table 1. Technical feature of the coupling.

Figure	Scheme	Characteristics [8]
	Spinning bicycle [9] + Generator	300W 15V 20A 2 800 rpm 85%

2.3. System configuration

The complete system is a group of 18 independent generating units, as shown in Figure 3.

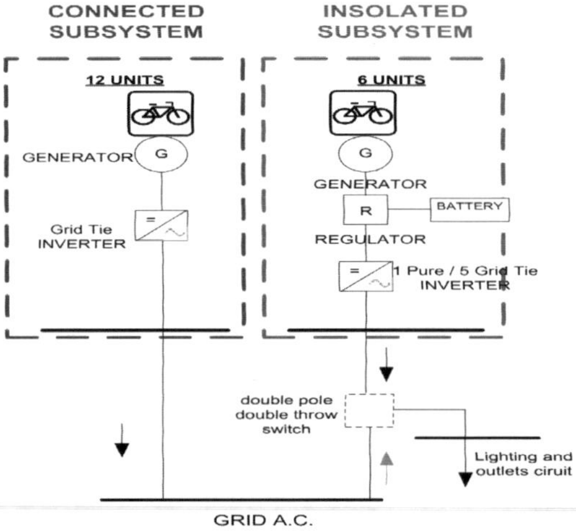

Figure 3. Basic diagram of the two subsystems.

The units of the first subsystem will inject real-time power to the grid, while the units of the second subsystem will supply power loads. All the settings will work independently, since it is not possible a parallel connection of generators, due to the variability in the speed of pedaling according to each user. The sizing and selection of conductors, ducts and protections were conducted under the guidelines of RETIE, NORMA ESSA and the provisions of the NTC 2050 [9] - [13].

2.4. Sizing the subsystem connected to the main grid.
Figure 4 shows the outline of a generating unit connected to the grid type.

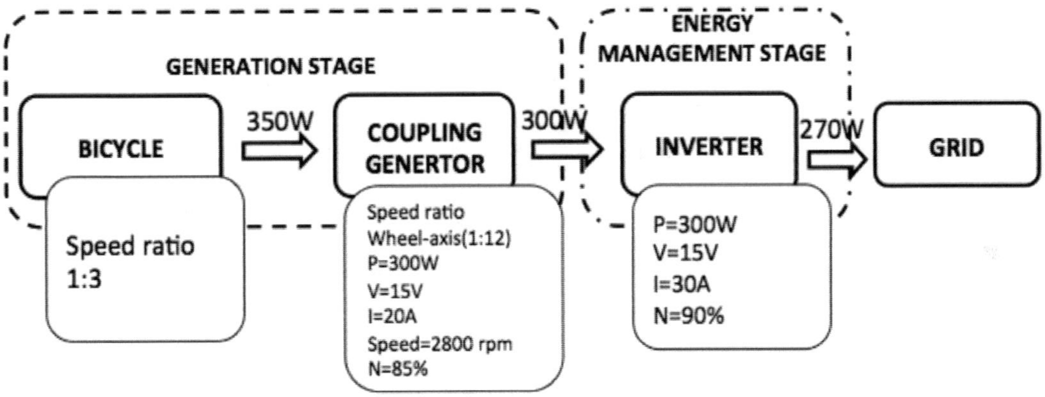

Figure 4. General diagram of the subsystem connected to the grid.

The management and power conditioning will be performed by a type *grid-tie* inverter with MPPT (Maximum Power Point Tracker), which will ensure the maximum possible power of extraction obtain at any time, since a smooth fit in the terminal voltage. Figure 5 shows the line diagram.

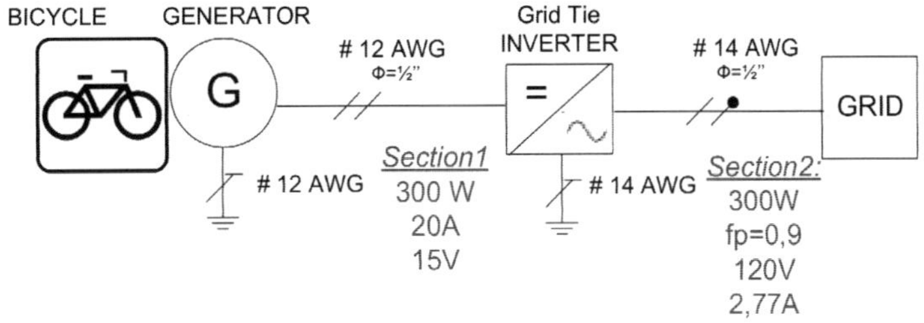

Figure 5. Line diagram of the generation unit – connection to the grid.

2.5. Sizing isolated subsystem
Figure 6 shows the nominal values of the components for each unit.

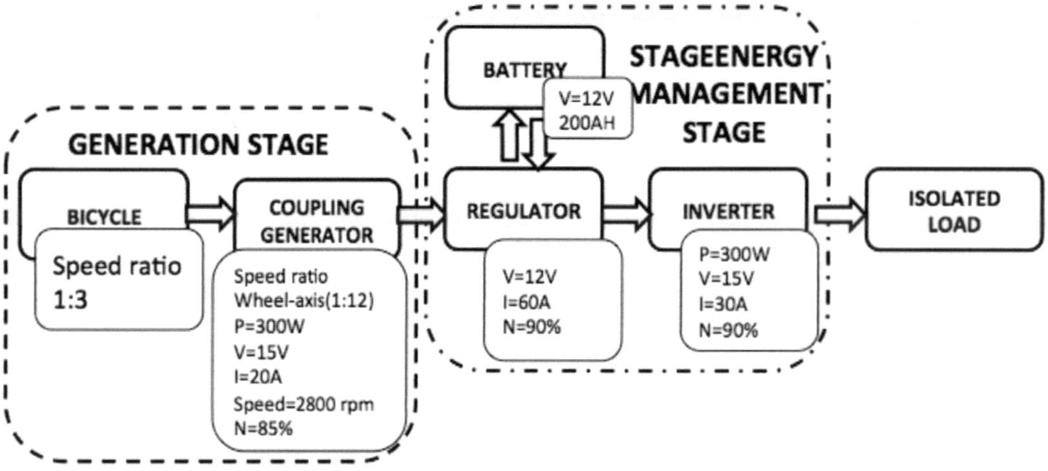

Figure 6. General scheme of the isolated subsystem.

The isolated subsystem must target specific power loads with an installed load of 1.8 kW from the energy generated in real time by spinning bikes and/or the energy stored on batteries. This power management subsystem will be made by regulators MPPT, inverter type Pure (to establish the conditions of operation: 120V and 60Hz) and five (5) Grid Ties (to be synchronized with the first inverter). The charge will supply the backup power grid in the building, which must be done by manual switching. Figure 7 shows the diagram.

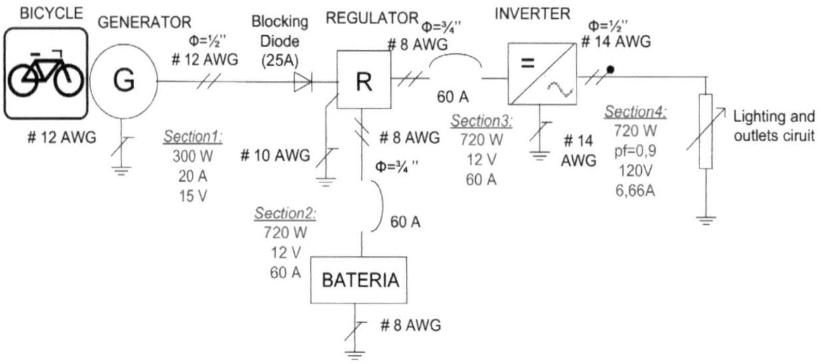

Figure 7. Direct driveline diagram with batteries.

3. Energy analysis

The following is an analysis of the energy behavior of the subsystem generator. Its purpose is to make an estimate of the power generation system.

3.1. Subsystem connected to grid

As presented in Figure 4, each unit will add 270 W to the grid approximately in average conditions (pedaling speed 2 800 rpm and a total efficiency of 76.5%). Therefore, it expects a surge of power of 3,2 kW for 100% occupancy.

3.2. Isolated subsystem

The analysis of this subsystem is more complex due to the battery charging and discharging. Therefore, we have six possible cases of operation per unit, as presented in Table 2.

Table 2. Cases of operation per unit in the isolated subsystem.

Case	Bicycle	*Battery*	Load	*Source*
1	Generate	Save	-	Bicycle
2	Generate	Save	>0	Bicycle
3	Generate	Full	>0	Bicycle
4	Generate	Delivery	>0	Bicycle + Battery
5	Generate	Save	>0	Grid
6	Not Generate	-	>0	Grid

In cases 1, 2, 4 and 5, the generator will operate at maximum efficiency, delivering the maximum power available at each moment, due to the system controller MPPT. If the batteries are fully charged (case 3) and there is an energy demand, the regulator will go off MPPT, and it will force the generator to operate less efficiently. In case of non-energy demand, the regulator will automatically stops its work, and the generator will open a circuit. When the unit (generator + battery) is unable to fill up the demand (cases 5 and 6), manual switching should be done to make the mains supply the request. In short, it means that the efficiency per unit will be 65.4%, in average conditions, and it will generate 230 W.

3.3. General behavior
In order to establish a representative analysis, it was assessed the behavior of 18 units during one-year period, considering two annual regimes: University schedule for a 22 days per a month for a 8 months-period and at the gym for a 30 days per month for a 12 months-period with two different types of daily operation (10 hours and 16 hours). Regimes and schedules are determined based on the dynamics of the use in the university and in the private gym. Table 3 shows the effective annual generation (real energy saving) per subsystem, unit, system and schedule.

Table 3. Estimated annual generation in MWh.

		Regime University		*Regime Private Gym*	
		10 hours	*16 hours*	*10 hours*	*16 hours*
Connection grid	Unit	0,47	0,76	0,97	1,56
	Subsystem	5,70	9,12	11,66	18,66
Isolated	Unit	0,41	0,65	0,83	1,32
	Subsystem	2,46	3,90	4,98	7,92
Total (MWh)		**8,16**	**13,02**	**16,64**	**26,58**

4. Design of a monitoring system of the energy behaviour

In order to know the energy performance, we designed a system that first monitors and displays variables, then, the data is analyze in a computational tool developed in National Instruments [14] called LabView. Due to its friendly interface to manage multiple visual signals and instrumentation at the same time. The visualization of the system behavior is possible for both staff and users. The set design has an intensive use of sensors and graphical display because it is a pilot case for research.

4.1. Data acquisition
For each unit connected to the sensor network are 5 and 9 sensors to isolated units, as they are shown in Figures 8, for a total of 114 signals.

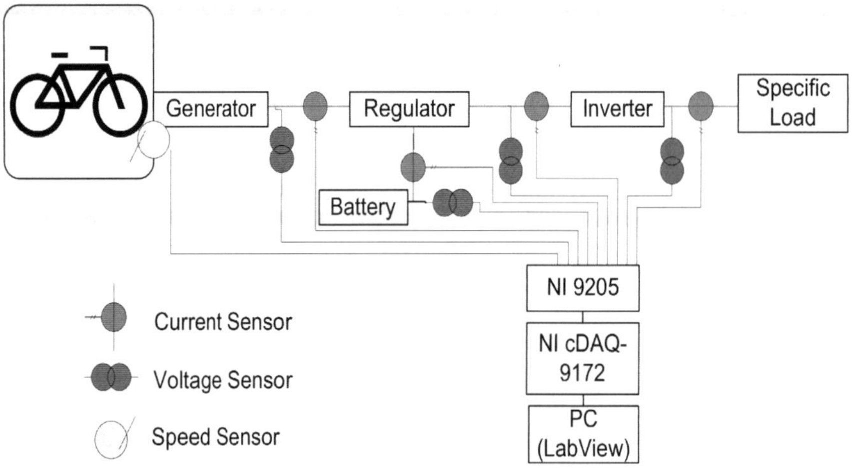

Figure 8. Block diagram of connection of the isolated subsystem.

To collect signals from the sensors 122, 8 of which are associated with injected and measured total power consumption of the network, we selected a data acquisition device called CompactDAQ NI-9178, with 8 acquisition module E / S OR 9205 of 16 entries.

4.2. Human machine interface

The interface consists of 7 tabs, which are presented in Table 4.

Table 4. Composition designed interface.

Tab	Visualized information	Beneficiary	Tool
A	Balance energético del subsistema conectado a la red	Technic personal	PC
B	Comportamiento detallado de cada unidad conectada a la red eléctrica	Technic personal	PC
C	Balance energético del subsistema aislado	Technic personal	PC
D	Comportamiento detallado de cada unidad aislada	Technic personal	PC
E	Balance energético del sistema	Technic personal	PC
F	Velocidad de pedaleo, de la potencia instantánea generada y la energía eléctrica generada durante la sección por usuario	User 1 al 9	TV LED
G		User 10 al 18	TV LED

The interface shows a warning, which is given in order to notify the staff that the when the isolated subsystem is insufficient to fulfill the energy demand for specific loads. This is detected when the voltage level or storage of all battery reaches its operating peak. The graphical interface requires an administrator to address the action and indicate the location of the bike user, which will be assigned automatically by the interface.

5. Financial analysis

For financial purposes, it was not taken into account the bikes price, their maintenance expenses. Therefore, the amount of investment to be considered is M $ 40. Table 5 determines the amount of investment required, the operation and maintenance costs, financial savings product harnessed the

IOP Conf. Series: Materials Science and Engineering **59** (2014) 012006 doi:10.1088/1757-899X/59/1/012006

power generated, along with financial indicators, NPV (net present value) and IRR (internal rate of return).

Table 5. Financial indicators.

Regime	Daily intensity (hours)	Annual saving USD	Average annual cost of operation and maintenance	VPN		TIR
				0%	5%	
University	10	1 430	1,05	-7 800	-9 900	-9,80%
	16	2 360		700	-3 350	0,70%
Gym	10	2 910		7 030	1 540	6,80%
	16	4 650		24 430	14 995	20,60%

The table shows that the invested money will be payed back after a ten-year period.

6. Results and Conclusions

Based on a test, it was concluded that a person will get an average pedaling speed closer to 242 [rpm], which has a potential energy close to 300 [W] at the output of a generator. It was designed a system of power generation of 5 [kW] of installed capacity, comprised of 18 independent generation units. The grid-connected units have 76,5% of efficiency and the isolated units have 68,8% of efficiency working without any connected battery. The total cost of the designed system is explained in the following table 6:

Table 6. Cost and efficiency per unit.

Units	Efficiency (%)	Lower cost (USD)
Grid-connected	76,5	1 200
Isolated	65,4	2 000

In short, the setting and management per unit will cost an amount of USD 1 000, which could generate between 0,4 MWh and 1,2 MWh per year, as a result it will save between $ 70 and $ 200. Additionally it requires an annual cost of operation and maintenance of approximately USD 35.

References

[1] Osma G 2011 *Uso racional de la energía en el edificio de Eléctrica II a partir del diseño de aplicaciones energéticas sostenibles* (Bucaramanga: Universidad Industrial de Santander)
[2] *¿Cómo es una bicicleta de spinning?* 2010 Retrieved November 6 2010 from www.foroindoor.com
[3] Osorio A 2007 *Generador eléctrico accionado por fuerza humana: una nueva alternativa de generación de energía* (Tabasco: Universidad Juárez Autónoma de Tabasco)
[4] Arango A and Fernández J 2011 *EKO-SPINNING* (Vitoria-Gasteiz: Instituto Politécnico Jesús Obrero)
[5] Strzelecki R, Jarnut M, Benysek G 2010 *Exercise bike powered electric generator for fitness club appliances* (Victoria:University of Zielona Góra)
[6] *Centro de Acondicionamiento físico BODYTECH* 2011 Retrieved July 17 2011 from www.bodytech.com.co
[7] Thomas L 1980 *The Analytical Hierarchical Process* (New York: University of Pittsburgh)
[8] Ramírez P 2010 *Aerogeneradores: Generadores de electricidad y productores de agua* (Quito: Universidad de Tacna)
[9] *System Pedal vision* 2011 Retrieved January 17 2011 from www.pedalpowergenerator.com
[10] *Catalogo HA Bicicletas* 2011 Retrieved January 20 2011 from en www.habicicletas.com
[11] Ministerio Educacion Nacional 2002 *NTC 2050 Primera actualización "Código Eléctrico*

Colombiano" (Bogota: Instituto Colombiano de Normas Técnicas y Certificación ICONTEC)

[12] Ministreio de Minas y Energía 2008 *RETIE reglamento tecnico colombiano de instalaciones electricas* (Bogota: Ministerio de Minas y Energía)

[13] Electrificadora de Santander 2005 S.A E.S.P. *Norma essa para el cálculo y diseño de sistemas de distribucion* (Bucaramanga: ESSA)

[14] *National Instruments*, 2011 Retrieved January 28 2011 from www.ni.co

IOP Conf. Series: Materials Science and Engineering **59** (2014) 012007 doi:10.1088/1757-899X/59/1/012007

Analysis of mismatch and shading effects in a photovoltaic array using different technologies.

J. Guerrero[1]*, Y Muñoz[2],F. Ibáñez[3] and A Ospino[4]
[1-2] Faculty of Energy Engineering, Universidad Autónoma de Bucaramanga UNAB, Bucaramanga, Santander, Colombia.
[3] Department of Electronic Engineering, Institute of Energy Engineering IIE Universidad Politécnica de Valencia UPV, Valencia, Spain.
[4] Renewable Energy, Universidad de la Costa CUC, Barranquilla, Colombia.

E-mail: jguerrero9@unab.edu.co

Abstract. In this paper, we analyze the performance of a photovoltaic array implemented in the Universidad Politécnica de Valencia which consists of modules of different technologies and power, connected in series, in order to quantify the energy losses due to mismatch and the effect of the shadows. To do this, the performance of the modules was measured in operation under ambient conditions with field measurement equipment (AMPROBE Solar Analyzer, Solar - 4000), which allows the extrapolation of measures to standard conditions STC. For the data validation, measures under controlled conditions were taken to some modules in the flash test laboratory of the Institute of Energy Technology ITE of Valencia in Spain. Subsequently the array curves measured were validated with a photovoltaic array model developed in MATLAB-Simulink for the same conditions and technologies. The results of this particular array are lost up to 20% of the energy supplied due to the modules mismatch. The study shows the curves and the energy loss due to shadows modules. This result opens scenarios for conceivable modifications to the PV field configurations today, chosen during the design stage and unchangeable during the operating stage; and gives greater importance to the energy loss by mismatch in the PV array.

1. Introduction

In recent years, with oil prices reaching record highs, nuclear plunged into crisis and international commitments to reduce emissions associated with conventional energy were made. Renewable energy has been gaining ground, and is seen to occupy a prominent place in the global generation, especially in the field of electric power. In this context, photovoltaic generation systems have the opportunity to be as much as suitable for their important advantage of being able to produce electrical energy very close to the electric loads. In this way the transmission losses are avoided and it is also possible to satisfy the daily load diagrams' peaks since they supply the maximum power quite in correspondence to the maximum request. The photovoltaic plants, moreover, do not emit pollutant emissions, do not vibrate and, thanks to their modularity, can comply with the morphology of the installation sites and so they present a lower environmental impact with respect to other renewable energy systems [1].

Optimizing the power output of a PV array is a very important task, especially in the novel Smart Grid context, that will rely on real time control and redirection of locally available power to loads [2].

One of the factors that affect the performance of a photovoltaic array is the mismatch. This term indicates the electrical maladjustment among the photovoltaic PV modules of the entire array. The causes of this maladjustment are attributable to the not homogeneous external characteristics of the

[1] *To whom any correspondence should be addressed.

modules, due to the dissymmetric manufacturing, the degradation of the module blooming layer, the manufacturing defects, the possible breaking of the cells, the dirt on the anterior part of the modules, the degradation of the materials used for the cells encapsulating, the unequal radiation of the modules, and the combination of different powers and technologies of modules in the array.

All these factors lead to a reduction of the array performances implying that the generated power of an array is less than the sum of the generated power for the single modules [2].

Several authors have analyzed the effects of using different power photovoltaic modules or mismatches, but using the same photovoltaic technology [3-5]. In this paper, we analyze the performance of a photovoltaic array, which consists of modules of different technologies and power, connected in series, in order to quantify the energy losses due to mismatch and the effect of the shadows.

2. Photovoltaic System

The photovoltaic system of Distributed Resources Laboratory (labDER) of the Universidad Politécnica de Valencia is an array of 11 photovoltaic modules of different technologies, powers and manufacturers as presented in Figure 1:

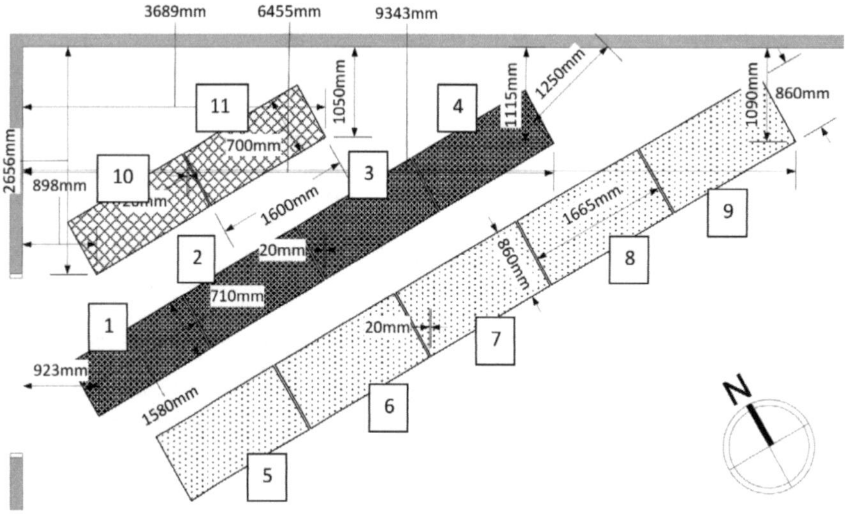

Figure 1. Photovoltaic field arrangement in the LabDER roof

- Modules 1 - 4
  Monocrystalline technology, manufacturer WANXIANG – China (180 W)

- Modules 5 - 9
  Monocrystalline technology, manufacturer REC – Norway (230 W)

- Modules 10, 11
  Polycrystalline technology, manufacturer USL – India (150 W)

The labDER photovoltaic system was studied to analyze the mismatch losses caused by the different panels' power and technologies.

2.1. Measurement System

Using an AMPROBE SOLAR – 4000 (Solar Analyzer) allowed to measure the characteristics of the static system installed in the LabDER roof. The power curves and power losses of each group of panels with the same type were obtained in outdoor conditions. In addition, to obtain the real curves of panels, the equipment generates the curves corresponding to standard conditions of measurement STC (irradiance of 1000 W/m^2 and temperature of 25° C) like a standard conditions in order to compare that results with other ones.

Every group of panels was measured independently and then the whole photovoltaic system was measured, getting I-V and power curves. In figure 4 it may be observed that I-V curve leans more to the left, indicating a lower efficiency of the USL panels compared to the other references.

The characteristics of the whole system are similar to the power curves of photovoltaic array with mismatch losses [4]. The effect of the low efficiency and power of the USL panels is a little bit remarkable in the power curve of the entire system. The goal of this work is to quantify these losses.

On the other hand, the STC power curve calculated for the solar analyzer is showed up for every array, resulting 658 W for the WANXIANG modules (Figure 2), 1082 W for the REC modules (Figure 3) and 253 W for the USL modules (Figure 4). A total power of 1993 W was expected, however the measured power results in 1585 W (Figure 5), obtaining a significant drop of 20% in the power of the global array.

These losses almost doubled those of the corresponding old and dirty panels [3], so we see that the use of panels of different technologies and manufacturers has much bigger influence.

- Measurement 1, Modules 1, 2, 3 y 4.

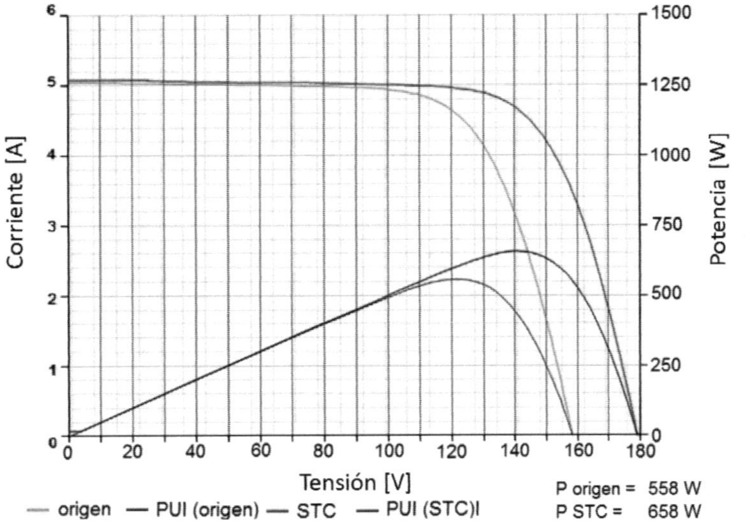

Figure 2. Curve I-V and P-V of WANXIANG panels

- Measurement 2, Modules 5, 6, 7, 8 y 9.

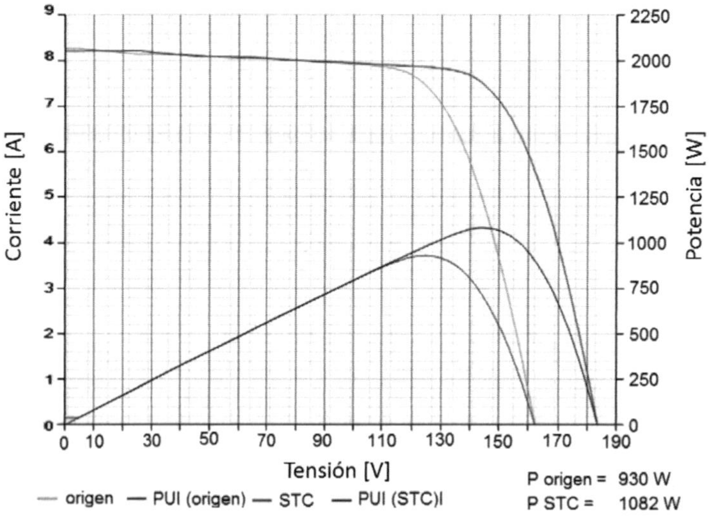

Figure 3. Curve I-V and P-V of REC panels

- Measurement 3, Modules 10 y 11.

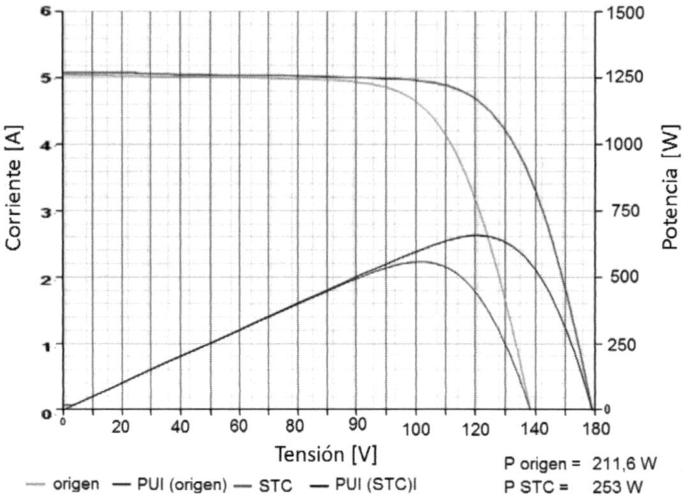

Figure 4. Curve I-V and P-V of USL panels

- Measurement 4, Total photovoltaic array

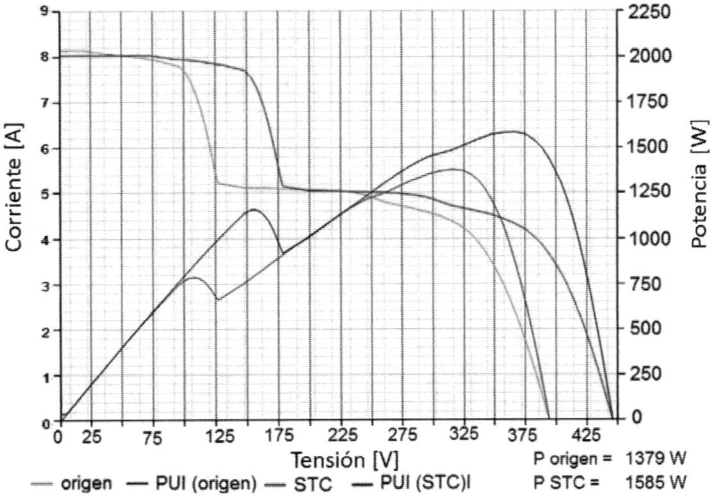

Figure 5. Curve I-V and P-V of total array

3. Verification of data reliability

As a part of the methodology of this study, an important item is the verification of the data provided by the field equipment AMPROBE SOLAR – 4000, because of the controversy that exists on the veracity of the data of in situ measurements. The results provided by the equipment AMPROBE are compared to those measured in the laboratory test of the photovoltaic panels, by means the "flash test", with this test, output values of V-I are measured in controlled irradiation and temperature conditions, obtaining the characteristic photovoltaic module curve. This test was made in the Institute of Electrical Technology ITE, this research center has the equipment, to obtain the electrical characteristics of the photovoltaic panels with a latest technology solar simulator, class A according to the standards IEC 60904-9, IEC 60904-1, and IEC 60891.

A test panel (Solar Plus – 40 W Module) was measured with both techniques (flash test simulator and field equipment), in order to evaluate if the results are correct. The measurement in the flash test (Figure 6) was 43,7W VS 41,81W obtained with the field equipment (Figure 7). The comparison gives an error of 4%, being according with similar studies published in the literature, which reports errors in the range of 3% - 4% when a field equipment is compared to the flash report [5].

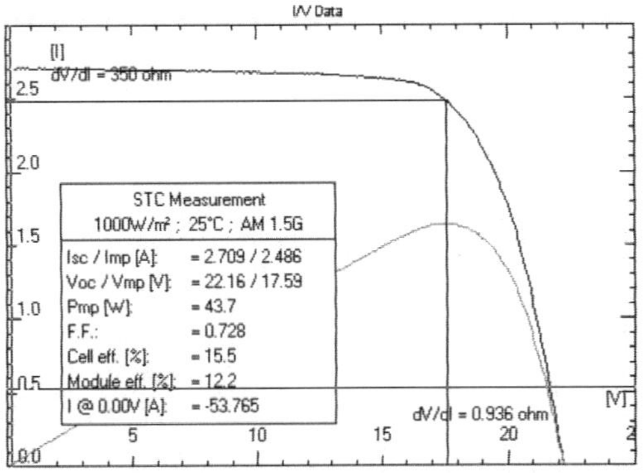

Figure 6. Curve I-V and P-V of flash report.

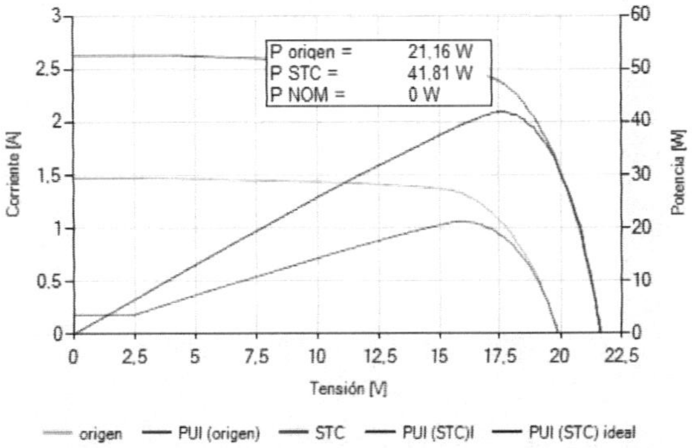

Figure 7. Curve I-V and P-V of Amprobe Solar.

Besides the used method to prove the veracity of field equipment with one of laboratory, there are other comparison methods as presented in [6].

4. Behavioral Validation

An important factor for ensuring the possibility of a power loss caused by the mismatch of about 20%, is to prove that this loss would not exist without mismatch. To validate the array´s behavior without mismatch, the LabDER photovoltaic generator was modelled, with the different types of modules connected in series.

The MATLAB-Simulink model is formed by 3 arrays (2*USL, 4*Wanxiang, 5*REC) connected in series, a load resistance and an electrical measurement equipment.

Every array models different type of modules and consists of a function block (Fcn), a controlled voltage source and conditional blocks (If, If Action Subsystem) as presented in Figure 8

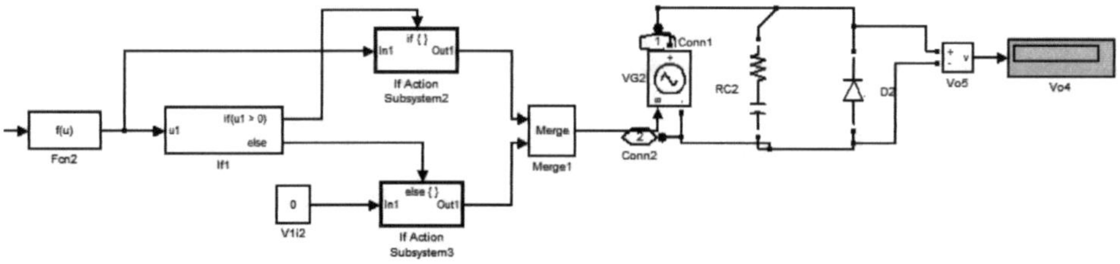

Figure 8. Simulink block model of a photovoltaic array of panels from the same manufacturer.

Instead of the typical I – V relation of the photovoltaic module, the function (Fcn) calculates the inverse function and gives a voltage dependent of the current panel and its characteristics, according to the next expression:

$$V_1(I_0) := N_{MS1} N_S \cdot V_t \cdot \ln\left(1 - \frac{I_0}{I_{SC1}}\right) + N_{MS1} \cdot V_{OC1} - I_{SC1} \cdot R_s \qquad (1)$$

The expression of the Fn1 Simulink block is:

$$u(7) * u(4) * u(5) * log(1 - (u(3)/u(2))) + u(7) * u(1) - u(2) * u(6) \qquad (2)$$

where:

u(1)= In1= Voc: Module Open Circuit
u(2)= In2= Isc: Module Short Circuit Current
u(3)= In3= Module output current
u(4)= N_S= 72: Number of cells of the module
u(5)= Vt= 26e-3: Thermal Voltage
u(6)= R_S: Module Series Resistance
u(7)= In4: Number of same type modules connected in series

Block Fcn output is a numerical value representing the voltage produced by the array formed by modules in series of the same type. In order to convert it into an electrical quantity, is applied to the control input of the controlled voltage generator.

IOP Conf. Series: Materials Science and Engineering **59** (2014) 012007 doi:10.1088/1757-899X/59/1/012007

The 3 blocks are connected in series and the output is applied to a resistive load. The current is the same in all the blocks and the output voltage is the sum of the voltages generated by each block. The characteristics curves I-V and P-V of the photovoltaic system are obtained varying the input current and measuring the load voltage and power.

Conditional blocks (If, If Action Subsystem) in order to avoid simulation error, producing no voltage if the current through a block is bigger than the short-circuit current. This fact happens by mismatch effect in systems with different types of modules, as in the present study; when a type of module produces a current bigger than the short-circuit current of another type, the bypass diodes connected in parallel with the less current module would drive, and the voltage through them would be almost zero. Figures 9 and 10 shows a power loss of about 20%.

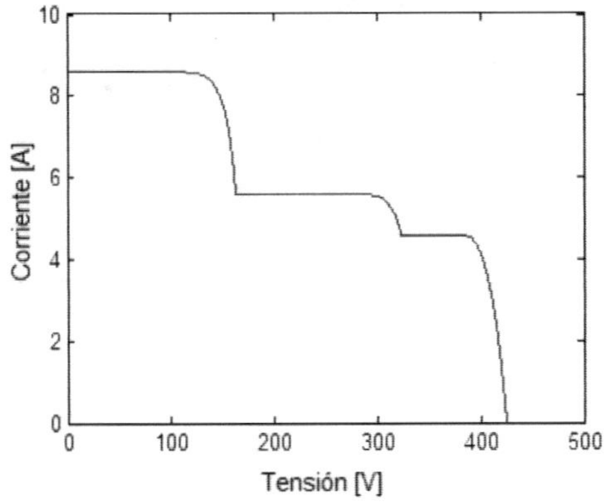

Figure 9. Curve I-V of Model Simulink.

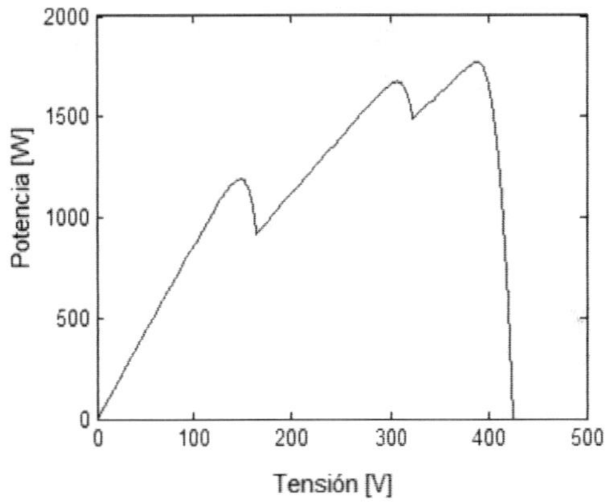

Figure 10. Curve P-V of Model Simulink.

This result agrees with the experimental measures, because without mismatch effect the system would produce a total power of about 1900 W.

The model could be modified to look for the optimum configuration, but these studies have already been reported in [1].

5. Shading Effects

The study of shading effects is important to foresee the working point of a system in case of shading. When a module or a part of it is shaded some of its cells become reverse biased, acting as loads instead of generators. If the system is not appropriately protected, hot-spot problem can arise and, in severe cases, the system can be irreversibly damaged [7]. In this section the losses caused by shadings in a photovoltaic module are analyzed, using a monocrystalline panel Solar Plus of 40 W. The process consists in shading a cell of the panel and measuring the characteristic curves before and after the shading effect. In addition pictures with a thermographic camera were taken, measuring temperatures in different points of the panel and relating the panel power loss and the temperature increase. The shading effect was made evident in the thermographic camera capture using a black paper with an emissivity = 1.

After shading a cell during several seconds the shade was removed, the results are shown in Figure 11, where is remarkable the heating of the shaded cell about to 55°C, also and a smaller heating of the next 3 chains of cells, caused by the change of the operating point of the shaded cell, that changes the behavior as a generator, to load.

Figure 12 shows the characteristic curves of the panel before and after the shading. Power loss is significantly high, the module reduces the power from 40,33 W to 3,935 W, almost a 90 % of power loss caused by the shading effects. These curves agree with the behavior of a shaded panel [8] but with a much bigger loss, even after several minutes a bigger heating in the nearby cells of the shaded cell is observed, that can causes the module damage. These results suggest future studies and advances in shading effects, because power losses are in the range of (40% - 90%), causing big cost in the photovoltaic field, considering is still an expensive technology.

Figure 11. Effect of a shaded cell in a photovoltaic panel heating different cells.

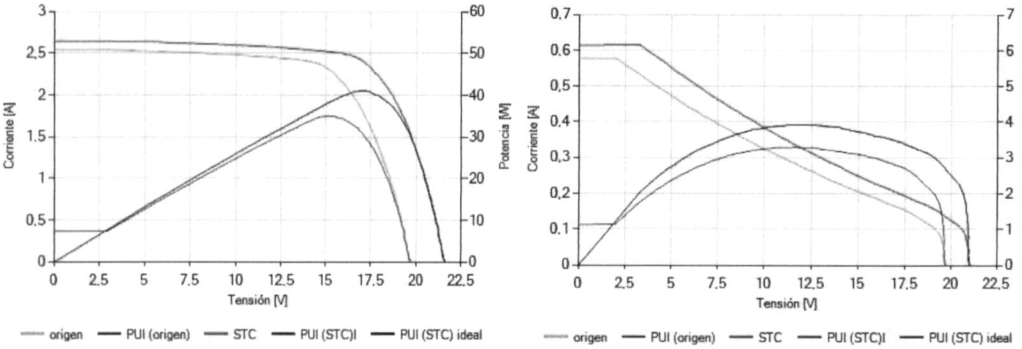

Figure 12. Power loss after shading a cell.

6. Conclusions

- The study shows an absolute error of 4% between the measuring with a field equipment vs laboratory measurements, an acceptable error in the engineering estimation area.

- Reference researches about mismatch losses in series connected arrays report bigger losses caused by ageing and dirt, however the present study prove that mismatch losses increase in array with different technologies.

- The MATLAB-Simulink model presented is compared to real operating curves of a photovoltaic array and successfully validate.

- This study proves that a photovoltaic panel can loss up to 90% of power by shading a unique cell. This shading causes heating in the shaded cell and in other cells because it becomes a load in the module.

7. References

[1] Dio V Cascia D Miceli R & Rando C, 2009. Clean Electrical Power, *International Conference on (Palermo: Italy/ Università degli Studi di Palermo)* **3**(8).
[2] Community Research European Commission. 2007. Strategic research agenda for Europe's electricity networks of the future *European Technology Platform Smartgrids* pp 12 - 26.

[3] Kaushika N & Rai A, 2007 An investigation of mismatch losses in solar photovoltaic cell networks *Energy* **32**(5) 755-759.

[4] Moschitta A Damiani A & Carbone P, 2012. A simple and accurate model for predicting mismatch effects in Photovoltaic Arrays *Energy Conference and Exhibition (ENERGYCON), 2012 IEEE International*. 818-822.

[5] Spertino F & Sumaili A J, 2009 Are Manufacturing IV Mismatch and Reverse Currents Key Factors in Large Photovoltaic Arrays?. *IEEE Transactions on Industrial Electronics*, **56**(11), 4520-4531.

[6] Wilson K Ceuster D & Sinton R A, 2006 Measuring the effect of cell mismatch on module output. In *Photovoltaic Energy Conversion, Conference Record of the 2006 IEEE 4th World Conference on* (**Vol. 1**, pp. 916-919).

[7] Garcia A Ruiz M C & Chenlo F, 2006 Experimental study of mismatch and shading effects in the I–V characteristic of a photovoltaic module. *Solar Energy Materials and Solar Cells*, **90**(3), 329-340.

[8] Garcia A Ruiz M C & Herrmann W, 2006 Computer simulation of shading effects in photovoltaic arrays *Renewable energy*, **31**(12).

IOP Conf. Series: Materials Science and Engineering **59** (2014) 012008 doi:10.1088/1757-899X/59/1/012008

Energy efficiency procedures for agricultural machinery used in onion cultivation (*Allium fistulosum*) as an alternative to reduce carbon emissions under the clean development mechanism at Aquitania (Colombia).

K Ochoa[1]*, S Carrillo[1] and L Gutierrez[1]

[1] School of Engineering, Cleaner Production Research Group (CHOC IZONE), El Bosque University, Carrera 9 No. 131 A - 02, Bogotá, Colombia.

Email: ochoakenneth@unbosque.edu.co

Abstract. Climate change has both causes and consequences over agriculture. This paper focuses on the first element and presents scenarios for ASOLAGO –an onion cropper's association in Colombia with 250 members– to reduce their carbon footprint. It evaluates a case study at "La Primavera" farm using a methodology approved by the United Nations Framework Convention on Climate Change. Land preparation and crop irrigation were analyzed as stages in order to propose energy efficiency alternatives for both the farm and the association. They include field efficiency, fuel economy and energy efficiency from biofuels for the first stage as well as solar and wind energy supply for the second. A cost-benefit analysis to generate additional income selling additional power produced by the system to the National Grid was done.

1. Introduction

This paper is the result of a research done for the most important onion (*Allium fistulosum*) farmers' association called –ASOLAGO–, located in Aquitania (Colombia). This municipality produces 75% of onion in Colombia representing 40% of the local income [1, 2].

In terms of environmental impact, Yoshikawa [3] analyzed carbon emissions related to this product through its lifecycle from extraction to final disposition. It was found a carbon footprint of 0,4 kg of CO_2 per kg of consumed onions.

The goals of this project were: i) to draw a baseline of carbon emissions at each crop stage, related to machinery use; and ii) to propose procedures using energy efficiency at selected stages.

The project was relevant to the guild, because of the lack of research on the product, and the increasing interest on two subjects: energy efficiency and clean development mechanism, at the agricultural level.

2. Methodology

The methodology used for this research was based on a deductive-descriptive analysis. It included the identification, registration and analysis of energy requirements for each step of the onion crop. As a reference, it was used the ninth version of the UNFCCC methodology "AMS II.F: Energy efficiency and fuel switching measures for agricultural facilities and activities" [1].

[1] *To whom any correspondence should be addressed.

2.1. Energy efficiency method selection
Two of the limitations that were considered to use this methodology were: i) geographical and physical location; and ii) total energy savings of the project are lower than 60 GWh per year. Three elements were considered to establish the baseline according to [1]: i) energy consumption should be reduced if adaptation actions are held; ii) if the energy is from fossil fuels, the baseline is equivalent to consumption for the activity; iii) it should be demonstrated the consumption of the agricultural activities, included the cultivated area and the crop performance.

This case study did not consider monitoring equipment transferred nor leaks that are present in them. Similarly, it was not applied to the same circumstances of a project activity under a program of activities specified in the methodology according to [1].

2.2. Methods used to calculate carbon emissions
In order to calculate carbon emission due to fossil fuels over the system, it was used the second version of the document provided by UNFCCC: "Tool to calculate project or leakage CO_2 emissions from fossil fuel combustion". To calculate carbon emission due to electricity consumption, it was used the first version of the document "Methodological tool to calculate project emissions from electricity consumption". It was used the emission factor for Colombia related to the National Interconnected System (SIN by its definition in Spanish) [2, 3, 4].

2.3. Methodology delimitation
Carbon emissions from soil preparation to onion yield were quantified. At "La Primavera", emissions related to energy use by agricultural machinery were calculated for soil preparation and irrigation, as shown on figure 1. Other cultural activities do not use machinery at the farm nor the association.

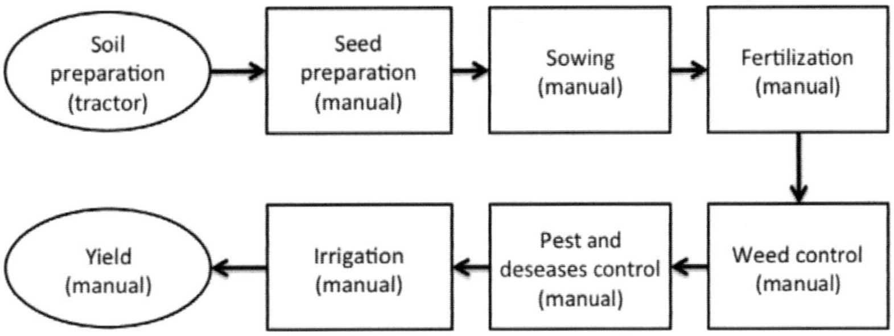

Figure 1. Onion production cycle

2.4. Other methodological considerations
One pilot survey was conducted to 25 farmers from the association to select the variables to be included on the final survey. The result was 25 variables classified into 5 categories: i) demographic information; ii) land characteristics; iii) soil preparation; iv) pests, weed and diseases control; and v) irrigation. Sampling unit for the survey was "the farmer". Sample universe correspond to 284 farmers. To determinate sample size, a 5% error continuous random variable was used, for a total of 58 farmers.

3. "La Primavera" baseline
Equation 1 was used to calculate carbon emissions at "La Primavera". Considering a cultivated area of 5.1 ha with an average of 3 harvests per year, 7.8 hours per ha for both plow and over-plow, and near 80 hours per harvest of soil preparation, carbon emissions correspond to 1.97 tCO_2/year. It was used fuel emission factor given by [5] as shown on table 1.

$$Em = FE \times H \times eff \tag{1}$$

Where,

Em	Are the total emissions in tCO_2 per year
FE	Is the energy of the fuel used in Gj
H	Is the number of harvests per year
eff	Is the emission factor of the fuel used in tons

A synthesis of the results is shown at table 1.

Table 1. Soil preparation associated emissions at La Primavera Farm

Variable		**Value**
Area (ha)		5.10
Harvests per year		3.00
Plow (h/ha)		7.80
Over plow (h/ha)		7.80
Soil preparation (h/harvest)		79.90
Diesel (gal/day)		20.00
Diesel consumption per harvest	(gal)	66.56
	(m^3)	0.25
	(kg)	209.73
Diesel energy (GJ)		8.85
Diesel emission factor (tons)		0.074
Carbon emissions per harvest (kg)		655.35
Total Carbon emissions (tCO_2/year)		1.97

Second stage –pump irrigation– used equation 2 to calculate carbon emissions. A total of 5.72 tCO_2/year are produced by this phase. Emission factor was provided by SIN. Results are shown on table 2.

$$Em = EC\ x\ eff \qquad (2)$$

Where,

Em	Are the total emissions in tCO_2 per year
EC	Electricity consumption per year in kWh
eff	Is the emission factor of the fuel used in tCO_2/kWh

The table 2 shows the results.

Table 2. Irrigation associated emissions

Variable	**Value**
Electricity consumption (kWh/month)	1,634.00
Emission factor (tCO_2/kWh)	0.29
Total carbon emissions (tCO_2/year)	5.72

Considering that just these two sources generate carbon emissions, total carbon emission was calculated by adding results from tables above. Table 3 presents the result and a 5 year and 10 year linear projection.

Table 3. Carbon emission scenarios

Term	Emissions (tCO_2)
1 year	7.69
5 years	38.45
10 years	76.90

4. ASOLAGO baseline

4.1. Characteristics of the association and its consumption patters

Results from "La Primavera" were used to forecast ASOLAGO's carbon emissions. Before showing results, it is important to highlight two characteristics of this organization: i) 8% of the owners are women and ii) 70% of the associates are over 50 years. Owners' lands have similar characteristics: 2.77 harvesting cycles per year, with 6 ha fields. 59% of the producers use tractors to do cultural activities with a 5.71 hours plow and 5.26 hours rake.

Tractors' cylinder capacity used by association members' determinate fuel consumption and therefore carbon emissions. 56% of farmers used low cylinder tractors (4,500cc or less), 44% medium machine cylinder capacity (4,500 to 9,000cc) and nobody large cylinder capacity engines (over 9,000cc).

Moreover, pest and diseases control is carried out by hand (cultural practices) and mechanized methods (motor sprayer). 29% of farmers use the second method, which is relevant to this study. Fuel consumption was estimated considering the average hours of the sprayer use; number of control activities per harvest cycle and average fuel consumption per sprayer. As a result, farmers consume 19.46 gallon of diesel per cycle.

For irrigation, farmers use water from Lake Tota and some water streams near their fields. 88% of farmers use electrical pumps, meanwhile 12% fuel pumps. For the first system, a theoretical demand of electricity was less than 3 MW, with an average use of 8.02 hours per day for irrigation during 1.5 days per week. For the second one, all fuel pumps use diesel. Three fourths of the farmers use electrical pumps with less and 47 hp. Figures 2 and 3 present a synthesis of the characteristics.

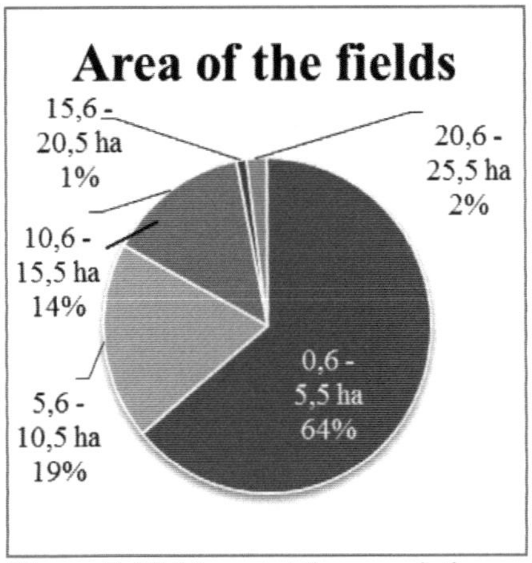

Figure 2. Field area at the association

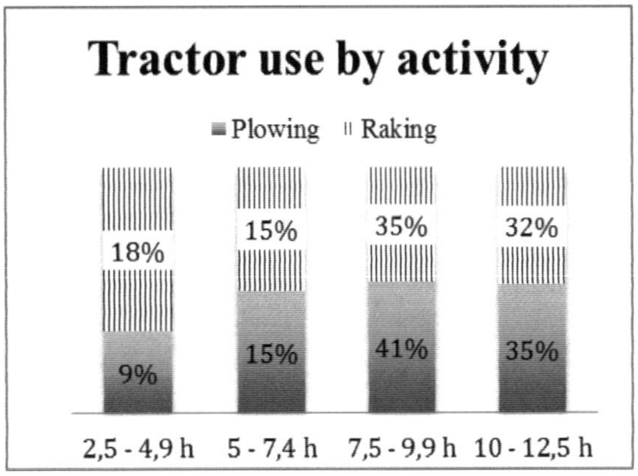

Figure 3. Tractor use by activity

4.2. ASOLAGO carbon emission

To calculate carbon emissions, information from a survey done to association was used. 88% of 250 farmers use electric pumps with an average consumption of 2,100 kWh per month. 56% of the farmers use similar tractors as the case study ("La Primavera"), used as reference for the association consumption considering that plowing and raking activities are similar.

Fumigation stage involves mechanical sprayers. Data from the survey was used as reference. It was found that 29% of producers use this type of equipment. In consequence, to calculate carbon emissions generated 82 of 284 producers were considered. In addition 2.35 gallon per day of fuel were used for pest control. Total emissions considering three categories are 2,190 tCO_2 per year. Results are shown in table 4.

Table 4. ASOLAGO carbon emissions

Variable	**Irrigation**	**Ploughing**	**Pest control**
Fuel consumption (GJ)	0	8.85	2.36
Electricity consumption (kWh/month)	2,100	0	0
Emission factor	0.2917	74.01	69.25
Carbon emissions per harvest (tCO_2)	0.613	0.655	0.163
Carbon emissions per farmer (tCO_2)	7.35	1.97	0.49
Farmers using machinery	250	159	82
Carbon emissions per activity (tCO_2 per year)	1,837.12	312.51	40.38
Total emissions (tCO_2 per year)		2,190.01	

5. Energy efficiency procedures to improve energy efficiency

In order to present energy efficiency procedures, two elements are presented for "La Primavera" farm: soil preparation using tractor and irrigation using pumps. Using this information a forecast for the association is done.

5.1. Soil preparation procedures using tractor

Soil preparation procedures were analyzed within two scenarios: i) field efficiency and fuel saving using theoretical values in comparison with real data; and ii, biofuels as an alternative to diesel.

5.1.1 Field efficiency and fuel saving. To improve soil preparation efficiency, it was necessary to consider the field theoretical capacity (FTC) and the fuel efficiency (FE). The first one is a relation between width work tool and theoretical speed, expressed in ha/h [6]. The second element is a theoretical value that corresponds to the efficiency. For this case it was used 74-90%, according to [7]. For the second variable, it is necessary to evaluate real versus theoretical field capacity. It is expressed as a percentage. For the farm it was found an efficiency of 12,2% for plowing and 8% for raking. Table 5 presents a synthesis of the information.

Table 5. "La Primavera" field efficiency

Operation	**Plowing**	**Raking**
Speed (km/h)	6 - 8	6 - 10
Efficiency (%)	74 - 90	74 - 90
Theoretical field efficiency (ha/h)	0.13	0.13
Real field efficiency (ha7h)	1.05	1.62
Field Efficiency (%)	12.20	8.00

In order to improve efficiency from both plowing and raking, it is necessary to evaluate different alternatives. First, according to [8], consumption varies by engine speed and load. Efficiency can be increased having an optimal use of the pedal and gearbox. This can be understood as an increase of the transformation of fuel into energy and consequently will improve machine efficiency. Theoretical analysis establishes that the best scenario for a 100 hp engine is 300 – 400 rpm, which will save 1.5 – 2 liters per hour [9]. Two scenarios are derived from these values. A low consumption one with a 3.71

gallons per hour at 1,500-2,000 rpm and 67-87 hp and an optimal consumption one with 3.6 gallons per hour at 1,500 rpm and 75 hp.

Plowing activities use 6 hour of machinery, which has 22.2-gallon consumption. Raking, on the other hand, use 4 hours of machinery that represents a diesel consumption of 14.8 gallon. Total consumption is 37 gallon per cycle and 111 gallon per year, considering 3 harvests during this term. In contrast, current fuel consumption is 200 gallons per year. This represents diesel savings of 89 gallon per year, equivalent to 1.09 tCO_2. This is a reduction of 0.88 tCO_2 emissions per year considering a 1.97 tCO_2 baseline.

Other best practices that can be implemented are related to engine maintenance. It includes cleaning air filters in order to reduce fuel consumption from 10 to 15%. Another procedure is the correct use of lubricants so as to protect engine from friction, heating and wearing [10].

5.1.2 Energy efficiency using biofuels. The second scenario was to improve energy efficiency using biofuels, considering that it is not necessary to have drastic changes over the engine [11,12]. It was selected B20 biodiesel, as it is the one Colombia has established since 2007 [13]. This generates a 1.5-tCO2 carbon emissions per year due to this fuel consumption. A 0.42 tCO_2 reduction has been calculated. This represents a 21.46% of emissions. In terms of energy, and considering recommendations by [14], a recalculation of the efficiency reports a 20.3%.

5.2. Irrigation procedures using pumps

For this element two scenarios were analyzed, based on general data on solar and wind potential of Aquitania, Boyacá. No specific analyses were performed due to time limitation.

5.2.1. Solar energy supply. In Boyacá, average solar radiation is between 4.5 to 5.o kWh/m^2. Solar brightness is from 5 to 6 hours of sun per day [15]. Pick power of the photovoltaic panel is 250 W with an area of 1.69 m^2. Each panel has a cost of 550 USD. Energy consumption demand is 37.3 kWh (50 hp). Whereas the system requires 12 hours of irrigation per day, during 4 days per month, total consumption by the pump is 1,790.4 kWh/month.

The cost, considering the government subsidy of the 50% [16] is approximately 2,200 USD per year. Table 6 presents a synthesis of the main elements within 5 scenarios of solar panels to cover 25%, 50%, 75% and 100% of the energy demand for the farm.

Table 6. Solar energy supply for irrigation at the farm

	Scenarios				
Element	**100%**	**75%**	**50%**	**25%**	**0%**
Energy generation (kWh/month)	13,428	10,071	6,714	3,357	0
Number of panels	504	378	252	126	0
Area (m^2)	298.0	223.5	149.0	74.5	0.0
Cost of panels (USD)	161,416.7	121,062.5	80,708.3	40,354.2	0.0
Financial profit	15,277.8	11,458.3	7,638.9	3,819.4	0.0
Electrical consumption (kWh/month)	0.0	447.6	895.2	1342.8	1,790.4
Carbon emission reduction (tCO_2)	5.72	4.29	2.86	1.43	0.00
Carbon emissions (tCO_2 / year)	0.00	1.43	2.86	4.29	5.72

The best scenario –100% of energy supply using solar panels– generates a 16,000 USD profit per year with a 9.3 year return of the investment (ROI) with 160,000 USD initial investment. This value is very high for one farmer. Another element to consider was required area, which correspond to 298 m^2. This was the reason to determinate a scenario for the whole association, which will be presented in the numeral 5.3.

5.2.2. Wind energy supply. Second scenario analyzed was to supply energy demand with wind energy. Here was important to consider local characteristics in order to design a system. To begin, wind speed in Boyaca is 3.1 m/s. Horizontal wind generator of 3.3 kWh with a low cost per Watt and a 20 years life cycle was used for calculations. Estimated cost was 6,555 USD per generator. Finally, potential generation is 117 kWh per generator in the farm [17, 18]. A synthesis of the results is shown in table 7.

The best scenario, 100% of energy supply using wind energy generators has a ROI of 47 years, and requires a 460 m^2 area for the project. 115 generators would be needed, creating problems such as landscape, noise and others, related to wind parks. This scenario is not attractive to the investors (local farmers) due to high costs.

Table 7. Wind energy supply for irrigation at the farm

	Installation scenarios				
Element	100%	75%	50%	25%	0%
Energy generation (kWh/month)	13,428	10,071	6,714	3,357	0
Number of generators	115	86	58	29	0.0
Area (m^2)	460.0	345.0	230.0	115.0	0.0
Cost of generators (USD)	752,376.1	564,282.1	376,188.0	188,094.0	0.0
Financial profit	15,254.6	11,440.9	7,627.3	3,813.6	0.0
Electrical consumption (kWh/month)	0.0	447.6	895.2	1,342.8	1,790.4
Carbon emission reduction (tCO_2)	5.72	4.29	2.86	1.43	0.00
Carbon emissions (tCO_2 / year)	0.00	1.43	2.86	4.29	5.72

5.3. Forecast for the association

A forecast was calculated for the association, taking into account variables of the local context. The first one was project design using 89% of the current associates, considering they use electrical pumps for irrigation. A synthesis is presented on tables 8 and 9.

Table 8. Solar energy supply for irrigation at the association

	Installation scenarios				
Element	**100%**	**75%**	**50%**	**25%**	**0%**
Energy generation (kWh/month)	3,021,300	2,265,975	1,132,988	283,247	-
Number of panels	67,140	50,355	25,178	6,294	-
Area (m^2)	113,467	85,100	42,550	10,638	-
Cost of Panels (USD)	36,367,500	27,275,625	13,637,813	3,409,453	-
Financial profit	3,430,957	2,573,218	1,286,609	321,652	-
Electrical consumption (kWh/month)	0.0	447.6	895.2	1,342.80	1,790.40
Carbon emission reduction (tCO_2)	1,837	1,378	919	459	-
Carbon emissions (tCO_2/ year)	0.00	459.28	918.56	1,377.84	1,837.12

Table 9. Wind energy supply for irrigation at the association

	Installation scenarios				
Element	**100%**	**75%**	**50%**	**25%**	**0%**
Energy generation (kWh/month)	3,021,300	2,265,975	1,510,650	755,325	-
Number of generators	25,823	19,367	12,912	6,456	-
Area (m^2)	103,292	77,469	51,646	25,823	-
Cost of generators (USD)	169,284,615	126,963,462	84,642,308	42,321,154	-
Financial profit	3,430,957	2,573,218	1,715,479	857,739	-
Electrical consumption (kWh/month)	0.00	447.60	895.20	1,342.80	1,790.40
Carbon emission reduction (tCO_2)	1,837	1,378	919	459	-
Carbon emissions (tCO_2 / year)	0.00	459.28	918.56	1,377.84	1,837.12

6. Conclusions

In conclusion, the AMS II.F methodology gave a first approach to calculate greenhouse gas (GHG) emissions to the project. However, if it is wanted to do a total balance of emissions in the cultivation of onions, this instrument does not allow carrying out this objective, which limits their use for future initiatives in the estimation of emissions.

The most relevant procedure for soil preparation at the farm is to change the fuel. This alternative does not imply additional investment and improves by 44.6% efficiency. For irrigation, solar energy is an environmental attractive alternative. However installation costs are very high for a single farmer. Initiatives like this must be accompanied of external investors (banks, multilateral funds, etc.). This could be reinforced with other alternatives such a clean development mechanism strategy. In order to improve this, it is important to prepare a project from the association.

Finally, project implementation related to alternative energies must evaluate financial dimension. This, because they require a very significant initial investment and such projects are very difficult to finance in contexts like small farmers at Colombia. On the other hand, to implement this kind of initiatives would benefit sectors like agriculture, which produce GHG at a local level. They could be part of the national inventories, which would give local governments the place at the agenda to attend cleaner production mechanisms and low carbon economies.

References

[1] United Nations Framework Convention on Climate Change, "Eficiencia energética y medidas de cambio de combustible para las instalaciones y actividades agrícolas," 2009.

[2] United Nations Framework Convention on Climate Change. (2008, August) Tool to calculate project or leakage CO2 emissions from fossil fuel combustion. [Online]. http://cdm.unfccc.int/methodologies/PAmethodologies/tools/am-tool-03-v2.pdf

[3] United Nations Framework Convention on Climate Change. (2007, June) Tool to calculate project emissions from electricity consumption. [Online]. http://cdm.unfccc.int/EB/032/eb32_repan10.pdf

[4] Haider Amaranto Sanjuán. (2010, August) Cálculo del factor de emisiones de CO_2 del sistema eléctrico interconctado para determinar la línea base de proyectos MDL. [Online]. http://www.siame.gov.co/Portals/0/Factor_CO2/Calculo%20del%20Factor%20de%20Emision_2009.pdf

[5] Academia Colombiana de Ciencias Exactas, Físicas y Naturales, "Factores de emisión de los combustibles colombianos," Bogotá, 2003. [Online]. http://www.siame.gov.co/siame/documentos/./mdl/DOC/18_FECOC.doc

[6] Santos Fino Alemán and Francisco Alvarez Aguilar. (2005) [Online]. http://es.scribd.com/doc/42792196/Manual-Maquinaria-Agricola-PDf

[7] Guido Botta. (2003) Licenciatura en negocios agropecuarios. Guía de clases Maquinaria Agrícola. [Online]. http://www.agro.unlpam.edu.ar/licenciatura/maquinariaag/Manual%20Lic.%20Adm.%20Costos%20Agropecuarios.pdf

[8] José Pérez de Ciriza and Alberto Lafarga. (2007, Enero) Navarra Agraria: Revista Técnica de Agricultura, Ganadería y Alimentación. [Online]. http://www.navarraagraria.com/n160/tractor3.pdf

[9] Khun S.A. (2012, Abril) Khun Seedliner. [Online]. http://www.kuhn-seedliner.com/internet/web_seedliner.nsf/0/8B534724749C5930C1257848004FA27A?OpenDocument

[10] Instituto para la Diversificación y Ahorro de la Energía. (2005, Abril) Ahorro de combustible en el tractor agrícola. [Online]. http://www.idae.es/index.php/mod.documentos/mem.descarga?file=/documentos_10255_Ahorro_combustible_tractor_agricola_05_a026b813.pdf

[11] John R. Agudelo, Pedro Benjumea, Edwin Gómez Meneses, and Juan Fernando Pérez Bayer,

"Biodiesel: una revisión del desempeño mecánico y ambiental," Ingeniería & Desarrollo, vol. 13, pp. 1-14, Enero 2003. [Online]. http://rcientificas.uninorte.edu.co/index.php/ingenieria/article/view/2320/1517

[12] Kubota Tractor Corporation. (2012, Abril) website of Kubota Tractor Corporation. [Online]. http://www.kubota.com/service/BioDiesel.aspx

[13] Unidad de Planeación Minero Energética de Colombia & Ministerio de Minas y Energía de Colombia. (2009, Abril) website of Unidad de Planeación Minero Energética. [Online]. http://www.upme.gov.co/Docs/Biocombustibles_Colombia.pdf

[14] Academia Colombiana de Ciencias Exactas, Físicas y Naturales, "Linea base para proyectos de sustitución de combustubles en la generación de energía eléctrica en zonas no interconectadas," Unidad de Planeación Minero Energética de Colombia, Bogotá, Informe Final Junio 2003. [Online]. http://www.siame.gov.co/siame/documentos/mdl/./7_LB_Leticia.htm

[15] Ministerio de Minas y Energía -Unidad de Planeación Minero Energética (UPME) y Ministerio de Ambiente, Vivienda y Desarrollo Territorial - Instituto de Hidrología, Meteorología y Estudios Ambientales (IDEAM), Atlas de radiación solar de Colombia. Bogotá, Colombia: UPME - IDEAM, 2005. [Online]. http://www.upme.gov.co/Docs/Atlas_Radiacion_Solar/1-Atlas_Radiacion_Solar.pdf

[16] EBSA S.A E.S.P. (2012, Enero) website of Empresa de Energía de Boyacá. [Online]. http://www.ebsa.com.co/wps/docs/NClientes/Enero_2012.pdf

[17] Ministerio de Minas y Energía de Colombia, Atlas de Viento y Energía Eólica de Colombia, UPME & IDEAM, Ed. Bogotá, Colombia: Imprenta Nacional de Colombia, 2006. [Online]. http://www.upme.gov.co/Atlas_Viento.htm

[18] Ecol-Energía S.A.S. (2012) [Online]. http://laenergia-solar.aquinegocio.co/p33285-aerogenerador-horizontal-para.html

Acknowledgments
Authors would like to recognize and thank the active participation of the Association of Onion Producers ASOLAGO of Aquitania, Boyaca (Colombia).

IOP Conf. Series: Materials Science and Engineering **59** (2014) 012009 doi:10.1088/1757-899X/59/1/012009

Experimental evaluation of energy efficiency in a gas-heated self-contained steam jacketed kettle.

V Manrique[1], L García[2]*, and J Alfonso[3]

[1-2-3]Technology Research Area, Gas Technological Development Center, Bucaramanga, Santander, Colombia

E-mail: lgarcia@cdtdegas.com

Abstract. Gas Technological Development Center (CDT de GAS) has developed a new *gas-heated self-contained steam jacketed kettle* as a solution for temperature-controlled cooking of pulp fruit in the Colombian traditional candy-making industry. This initiative seeks to contribute to the promotion of natural gas as a cleaner and more efficient alternative to traditional fuels such as wood and coal in the Andean countries. Prototype follows the operational and safety rules provided by international and local standards: ASTM-F1602, ASME (section VIII, division 1), and NTC-4082. This paper presents the methodology, experimental setup and results obtained during the performance tests for *heating efficiency* evaluation of the kettle, according to the ASTM-F1785 standard, and the estimation of measurement uncertainty according to GUM method. The *heating efficiency* rate for this new prototype is higher to the convectional and commercial models of gas heated self-contained steam jacketed kettle. The results showed a heating efficiency of 63.83% ± 1.66% (k=2), whereas the emission of CO and NO_X in stack gases was under the regulate limits for natural gas equipment.

1. Introduction

Traditional production of fruit candies is an important regional economic activity in Colombia. Hundreds of small and family businesses depend on this economic activity. For example, annual production of "bocadillo" of guava in Velez (Santander) and Ricaurte (Boyacá) regions amounts to some 24,000 to 35,000 ton a year, with a turnover of USD 24 million. Another example is the production of "arequipe" in Floridablanca (Santander) and "Manjar Blanco" in Valle del Cauca [1].

Basically, the production process consists of increasing the soluble solids concentration (Brix degree) by adding sugar (saccharose) and evaporating the water from the fruit pulp at a controlled temperature [2].

In the last decade this sector has been modernized by the introduction of best manufacturing practices (BMP) and the emergence of new lines of products and packages intended to reach international markets. Some technological improvements include the introduction of electric pulpers and the substitution of copper pots with stainless steel jacketed kettles whose steam supply comes from a steam boiler, usually fired by solid fuels such as coal or wood, because of their relatively lower price and its abundance in the region [3].

[2] *To whom any correspondence should be addressed.

These advances have faced cultural and economic barriers hindering the sustainable development of this traditional sector. Public health issues, industrial safety and local air pollution are still of concern [4]. The Gas Technological Development Center (CDT de GAS in Spanish) seeks to contribute to the promotion of natural gas as a cleaner, safer and more efficient alternative to traditional fuels such as wood and coal. With that purpose, it is currently developing technological solutions for the sustainable use of natural gas in the traditional sectors described above.

2. Direct steam vs. gas-heated self-contained steam jacketed kettles

Direct steam kettles are semi-spherical-shaped containers, with a double bottom known as jacket. A heating fluid, e.g. water steam, flows through the jacket. The product is heated at steam saturation temperature, related to operating pressure in the jacket, typically 207 kPa - 276 kPa. The steam condenses and is drained off. Occasionally, this water is sent back to the boiler for a new cycle. Steam generation, distribution (pipes, control valves, etc.) and water return (pipes, pumps, valves, etc.) to the boiler demand extensive infrastructure under this production scheme.

Self-contained kettles, meanwhile, do not depend on an external boiler. The jacket contains water and an attached heating system (electric or gas). Steam is generated inside the jacket. Steam heats the product at the saturation temperature related to the operating pressure, but each time the condensed water drops into the jacket and boils again. In this case, the infrastructure for steam distribution and condensation return is not required [5].

This scheme is an *innovative process* vis-à-vis the traditional process, which relies on a boiler and a direct-steam kettle. Additionally, replacing coal and wood with natural gas has environmental, safety and operational benefits.

Since 2010 CDT de GAS, financially supported by the Colombian state science department COLCIENCIAS, has worked on the development of two self-contained steam jacketed kettles prototypes. The most important performance parameter is *energy efficiency*, since it is directly related to fuel costs [6]. These prototypes take into account the safety, control and performance requirements provided by ASTM F1602 [7], ASME Boiler & Pressure Vessel Code [8] and related Colombian Technical Standards (NTC in Spanish), such as NTC-4082 [9].

In 1997, a new ASTM standard for steam kettles, designed as *ASTM F1785-97* [10]*: Standard Test Method for Performance of Steam Kettles,* was developed in the United States. This is a direct method that allows to evaluate and report the maximum energy consumption (*firing rate*), production capacity, heating efficiency and simmer energy rate [5]. The results, according to ASTM F1785 performance evaluation for the latest prototype developed by CDT de GAS, are shown hereafter.

3. Heating efficiency according with ASTM F1785

ASTM F1785 is a direct method for determining heating efficiency, as the ratio between the useful energy (sensible heat) and the supplied energy (chemical energy in the natural gas). Supplied energy is calculated from the natural gas volume (at standard conditions) and the High Heating Value. According to this procedure, the kettle pot is filled with clean water to 90 ±1% of the fill-to-spill volume. Water temperature is increased from 26.67 ℃ to 71.12 ℃ while natural gas pressure, temperature, volume and time are measured [10]. The mathematical expression for heating efficiency is:

$$\eta = \frac{\text{Useful Energy}}{\text{Supplied Energy}} = \frac{m \cdot C_p \cdot (T_f - T_i)}{V_{std} \cdot H_V} \qquad (1)$$

Where m is the water mass, C_p is the specific heat of water, $(T_f - T_i)$ is the temperature increasing, V_{std} is the natural gas volume at standard conditions, and H_V is the High Heating Value for natural gas.

3.1. Test conditions

The performance test for the self-contained steam jacketed kettle was developed using instruments and equipment from CDT de GAS. These instruments were selected following the appropriate range and metrological characteristics, including gas, temperature and volume of natural gas, mass and temperature of water. CDT de GAS Metrology Center is accredited by the National Accreditation

IOP Conf. Series: Materials Science and Engineering **59** (2014) 012009 doi:10.1088/1757-899X/59/1/012009

Agency of Colombia (ONAC in Spanish), according to ISO/IEC 17025:2005 [11] (Accreditation code: 10-LAB-013 [12]).

Figure 1 shows the schematic assembly for the test, and Table 1 shows the metrological characteristics of the instruments.

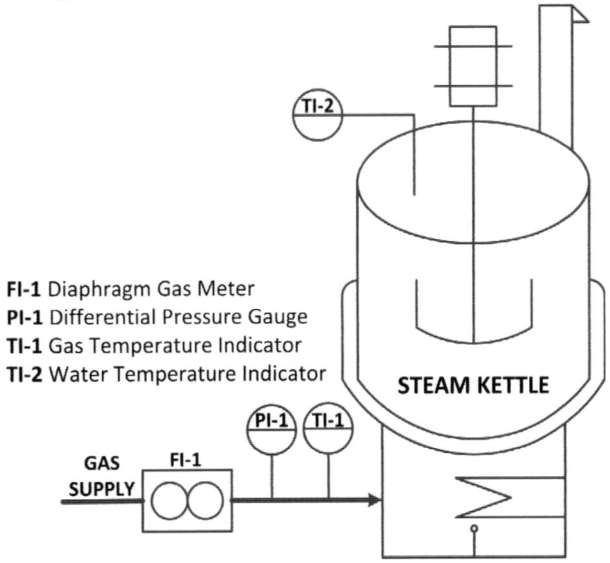

Figure 1. Schematic assembly for the test.

Table 1. Measurement Instruments.

Instrument	Range	Resolution
Scale	0 kg - 260 kg	0.05 kg
Barometer	81.1kPa - 106.1 kPa	1 Pa
Diaphragm Flow Meter	0.025 m^3/h - 4.0 m^3/h	0.001m^3
Differential pressure gauge	0 kPa -5 kPa	0.03 kPa
Temperature sensor	0 °C – 120 °C	0.1 °C
Chronometer	0- (9h;59m;59.99s)	0.001 s

Natural gas used in the performance test is produced in *Gibraltar Well*, and supplied by natural gas transmission company TRANSORIENTE SA ESP. The composition and High Heating Value was evaluated by chromatography in the CDT de GAS Metrology Center following ASTM D3588 [13]. Table 2 shows the properties of the natural gas, and Table 3 summarizes other specific test conditions.

Table 2. Fuel gas properties[a]

Standard High Heating Value, ASTM D 3588		40.13 MJ/m^3
Standard Density, ASTM D3588		0.768 kg/m^3
Standard Compressibility Factor, AGA 8		0.9975
COMPOSITION OF NATURAL GAS GIBRALTAR (ASTM D1945 [14])		
Component	Formula	Molar concentration, % mol
Nitrogen	N_2	0.528
Methane	CH_4	89.178
Carbon dioxide	CO_2	1.850
Ethane	C_2H_6	5.759
Propane	C_3H_8	1.807
Iso-butane	C_4H_{10}	0.404
n-Butane	C_4H_{10}	0.341
Iso-pentane	C_5H_{12}	0.063
n-Pentane	C_5H_{12}	0.041

n-Hexane	C_6H_{14}	0.029

[a] Analysis Report INRA-12-CRM-013-1951. Gas Quality Laboratory. CDT de GAS. July 19th 2012.

Table 3. Test conditions

Variable	Value
Average Atmospheric Pressure	90.5 kPa
Average Local Temperature	299.2 K
Water (mass)	85.0 kg
Water (Specific Heat)[b]	4.18 kJ/(kg·K)

[b] Calculated using NIST Reference fluid thermodynamic and transport properties database - REFPROP

3.2. Results

Table 4 shows the results of the last 3 tests. However, 10 prior tests were carried out in order to tune and optimize the water level for boiling, and to estimate the correct position of gas injectors.

Table 4. Test Results

Variable	Test 1	Test 2	Test 3	Average
Injectors orifice (mm)	2.38	2.38	2.38	2.38
Jacket vaporization water (dm^3)	18	18	18	18
Gas supply pressure (kPa)	1.47	1.37	1.39	1.41
Gas supply temperature (K)	300.7	300.7	300.7	300.7
Gas volume at supply conditions (dm^3)	637	635	636	636.0
Gas volume at standard conditions (dm^3)	556.29	553.94	554.96	555.07
Test time (s)	1400	1432	1432	1421.3
Supplied Energy rate (kW)	15.95	15.54	15.55	15.67
Useful Energy rate (kW)	10.16	9.93	9.93	10.0
Heating efficiency (%)	63.69	63.96	63.85	63.83

Heating efficiency was 63.83%, this a high efficiency according with previous benchmarking studies.

4. Uncertainty

ASTM F1785 suggests a method for uncertainty estimation, as the product of the Standard Deviation and an Uncertainty Factor, which depends on the number of results. The Standard deviation shows the dispersion from the representative value (arithmetic average). Accordingly, standard deviation shows the repeatability of the test, but it does not take into account the uncertainty due to metrological performance of instruments [10].

4.1. Uncertainty calculated through ASTM F1785

Uncertainty for heating efficiency was estimated through the ASTM F1785 procedure. The arithmetic average for 3 results in Table 4 is 63.83 and the Standard Deviation is 0.136. For 3 results, the Uncertainty Factor is 2.48 (taken of Table A1.1 ASTM 1785 [10]). So Absolute Uncertainty is equal to the product of 0.136 and 2.48, 0.337. Finally, the expression of heating value should be (63.83 ± 0.337) % or, in relative terms, 0.638 ± 0.53%.

4.2. Uncertainty calculated through Propagation of Uncertainty

There are uncertainty sources that are not considered in the ASTM F1785 procedure. Figure 2 shows a schematic cause-and-effect diagram with the uncertainty sources related to each parameter included into the mathematical function of the heating efficiency.

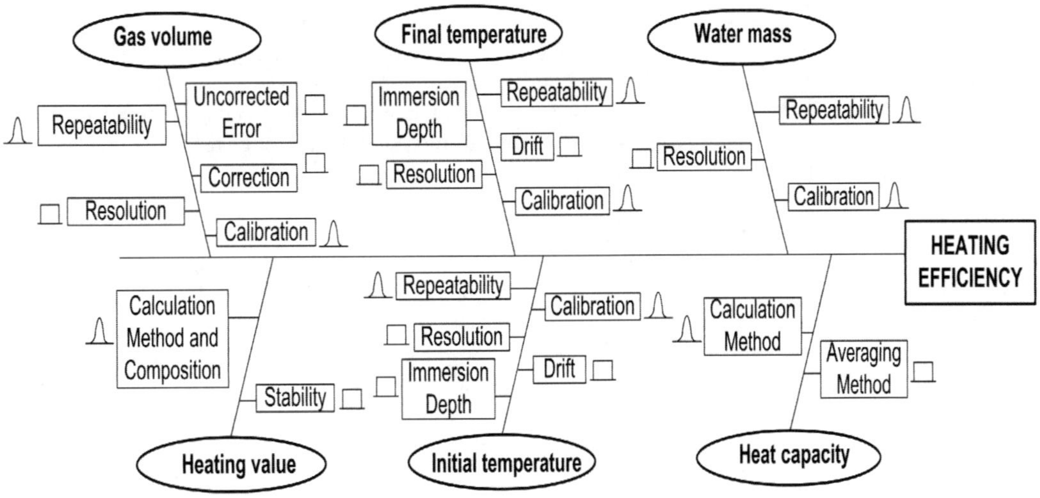

Figure 2. Uncertainty sources of the heating efficiency function

As showed above, heating efficiency is a function of pressure, temperature, volume, heating value, specific heat, and mass. Each variable has uncertainties due to measurement instrument limitations or the quality of data acquired. Probability function associated to each uncertainty source can be a normal or rectangular distribution according to the data evaluation.

For example, the resolution of the scale used to measure the mass of water, limits the lower change in the mass of water that is possible to measure. Uncertainties due to repeatability refer to standard deviation of experimental values measured for a variable. Calibration uncertainties are derived by a calibration report for each instrument. Drift uncertainties refers to the stability of the instrument in a time line between calibrations. Uncertainties of High Heating Value of natural gas and Specific Heat of water depend on the calculation method.

Propagation of Uncertainty method takes into account the effect of each uncertainty variable on the result of the uncertainty function. Due to the statistical probability distribution of each variable that is known or can be assumed, it is possible to derive confidence limits to describe the region within the true value of the variable to be found [15]. Table 5 shows the uncertainties for each variable and the results obtained, according to (1).

Table 5. Uncertainty Budget for heating efficiency measurement

Input magnitude Xi	Input estimate x_i	Standard Uncertainty $u(x_i)$	Sensitivity coefficients C_i	Contribution $C_i \cdot u(xi)$	Contribution (%)	Degrees of freedom v_i
Water mass (kg)	85.00	0.0645	0.00751	0.00048	0.81	3
Heat capacity (J/kg·K)	4182.6	3.6	0.00015	0.00055	1.05	149
Final temperature (K)	70.0	0.18	0.01596	0.00280	26.95	66
Initial temperature (K)	30.0	0.18	-0.01596	-0.00280	26.95	66
Gas volume (m^3)	0.55507	0.00202	1.15002	0.00232	18.48	66
Heating value (J/m^3)	40135597	172093	0.00000	0.00274	25.77	123
Combined standard uncertainty, $u_c(y)$		0.0054				
Effective degrees of freedom, V_{eff}		304				
Coverage factor, k		1.97				
Expanded uncertainty, U		0.0106				
Relative uncertainty, U_{rel}		1.66%				
UNCERTAINTY REPORT						
Heating Efficiency		0.638	± 1.66%			

IOP Conf. Series: Materials Science and Engineering **59** (2014) 012009 doi:10.1088/1757-899X/59/1/012009

Finally, heating efficiency was calculated as (63.83 ± 1.06) % with a confidence interval of 95.5% (in relative terms: 0.638 ± 1.66%). This uncertainty estimated following the *"propagation of uncertainty"* method is higher than the uncertainty estimated through ASTM F1785. As explained earlier, uncertainty is higher due to the additional sources taken into account, but it provides a better outlook on the source of uncertainty and the best method to reduce the effect of the instruments on the measured results in order to increase the reliability of future tests.

5. Boiler Efficiency evaluated through a gas analyzer

Boiler efficiency assesses the use of the released energy in the combustion, and is equal to the product of the *combustion efficiency* and the *heat transfer efficiency* from flue gas to boiling water.

Combustion efficiency represents the ratio between the released energy by combustion and available energy in the fuel. This efficiency depends on the technology of the burner as well as the shape and size of the combustion chamber and the air/fuel ratio. Typically, the air excess must be 10-20%, because deficiency of oxygen (contained in air) produces high carbon monoxide (CO) emissions. High excess of air and high temperatures produce nitrogen oxides (NOx) emissions.

High excess of air produces a high volume of inert products (as nitrogen) in the flue stack. The excess of flue gas reduces the heat available for transfer to vaporization water in the heat exchanger, as part of this energy is removed by the excess of nitrogen in flue gas [16].

Boiler efficiency can be evaluated with a *Flue Gas Analyzer*. This equipment measures the concentration of oxygen, carbon dioxide, carbon monoxide, nitrogen oxides and flue stack temperature. With these parameters, the Analyzer calculates the air excess in the flue gases, and the *boiler efficiency* of the self-contained steam jacketed kettle [17]. Table 6 shows the metrological characteristics of the Flue Gas Analyzer.

Table 6. Metrological performance of Flue Stack gas Analyzer [18]

Device	Gas Handheld Analyzer AMPRO 2000	
Manufacturer	MRU Instruments, Inc.	
Component	Range	Accuracy
O_2 Oxygen	0-21 Vol-% abs.	± 0.2 Vol-% abs.
CO_2 Carbon dioxide	0 % -20 %	± 0.4 Vol-% abs.
CO Carbon monoxide	0ppm -4000 ppm	± 20 ppm o 5% reading < 4000ppm
NO Nytrogen monoxide	0ppm -1000 ppm	± 5 ppm o 5% reading <1000ppm
NO_2 Nytrogen dioxide	0ppm -200 ppm	± 5 ppm o 5% reading <200ppm
T-Gas Gas flue temperature	≤ 650 °C	± 2.0°C <200°F/1% reading >200°F
T-Air Local Air temperature	≤ 100°F	± 1.8°C
Boiler Efficiency	0 % - 100%	-
Air excess	0 % - 99.9 %	-

Boiler efficiency was measured at the same conditions of the heating efficiency test. According to calculations made by the analyzer, excess air was set between 15-20%. Boiler efficiency decreased from 89% to 84% as flue stack temperature increased from 164°C to 210°C. Figure 3 shows this behavior: boiler efficiency is inversely proportional to flue stack temperature, because the temperature difference between the combustion gases and the vaporization water decreases during the test, so the heat transfer potential is lower at the end of the test.

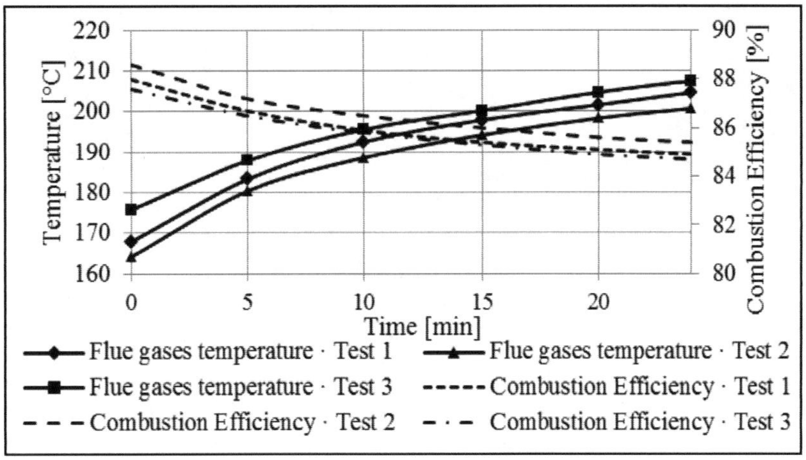

Figure 3. Transient Combustion Efficiency

6. Simmer Energy Rate

ASTM F1785 established a procedure to determine the simmer energy rate. It accounts for the firing energy rate necessary to hold the cooking product at set temperature for a period of time. This test is conducted after the heating efficiency test. The kettle temperature controller is set to 74 ± 0.5°C. When the kettle reaches this temperature, the controller shuts off the main solenoid valve, and the gas burners turn off. When temperature is below the set point, the controller opens the solenoid valve, the burners turn on, and temperature increases again until the set point is reached. The simmer energy test was conducted for 3 hours, and several cycles of heating (burners being turned-on) and cooling (burners being turned-off) may occur during this time span. [10]

Figure 4 shows the temperature behavior during the test. Set-point temperature was 74°C and the temperature controller had a minimum hysteresis of 1°C, so the upper limit was 74.5°C and the lower limit was 73.5°C.

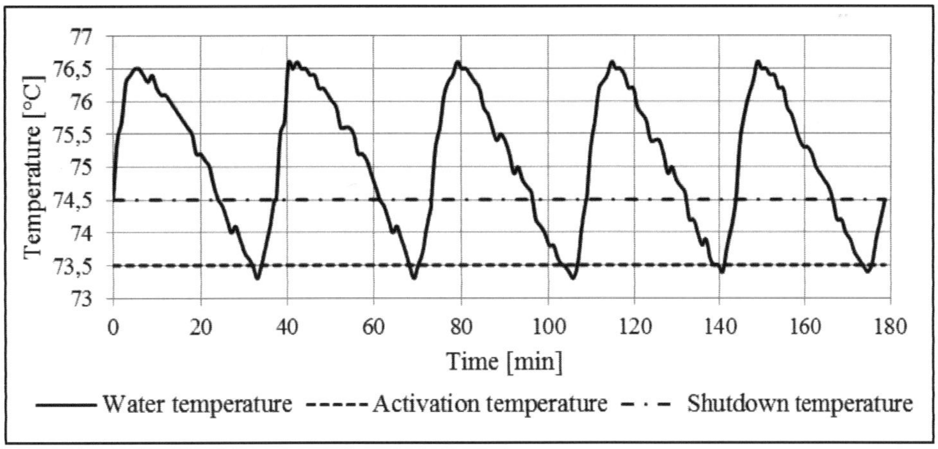

Figure 4. Simmer Energy Rate Test

Water temperature changed from 73.3 °C to 76.6°C. Although the burners are turned off when temperature reaches 74.5°C, temperature increases to 76.6°C due to accumulated energy as steam pressure (latent heat). When temperature reaches the peak, it decreases as the internal pressure decreases due to boiling down of steam inside the jacket. It is important to remember that steam temperature (and cooking water) is directly related to and depend on the saturation pressure (operating pressure inside the jacket). Heating time (burners turned on) lasts 5 minutes, and cooling time (burners turned off) lasts 30 minutes, so keeping a given temperature requires that kettle stays off longer than it stays on, reducing fuel costs.

Simmer energy rate is calculated as the ratio between the net volume of natural gas and the test time (approx. 3 hours). In this case, the apparent energy rate is 2.19kW. Power is just 13% of the nominal energy rate for heating.

7. Conclusions

It was applied the method of *"propagation of uncertainty*, as an alternative to the method proposed by the ASTM F1785, whereby a more reliable estimation of the results was obtained during test heating efficiency.

Energy efficiency of the self-contained steam jacketed kettle was assessed according to ASTM F1785 and two representative parameters were evaluated: (a) heating efficiency, with average value of 63.83±1.06% for a nominal firing rate of 15kW, and (b) simmer energy rate, with average value of 2.19kW.

The Boiler efficiency, for the new kettle, was measured at the same conditions of the heating efficiency test. With an excess air between 15%-20% the Boiler efficiency varied from 89% to 84% as flue stack temperature increased from 164°C to 210°C.

The performance is better than similar appliances. According to benchmarking published by the United States Food Service Technology Center (see the *Steam Kettle Technology Assessment* report [5]), same-size conventional self-contained kettles (gas-heated) with a similar size have a heating efficiency ranging between 39%- to 54%, as evaluated with ASTM F1785.

8. References

[1] Rodríguez Borray G 2005 *Estudio del Sistema agroalimentario Localizado SIAL, de la concentración de fábricas de bocadillo de guayaba en las Provincias de Vélez y Ricaurte en Colombia* (Corpoica)

[2] Burgos N, Zárate K and Escalante H 2006 *Diseño conceptual del proceso productivo de dulces de guayaba utilizando gas natural en el municipio de Moniquirá* Proyecto de Grado (Bucaramanga, Santander: UIS-Ingeniería Química)

[3] Mesa Gestora: Cadena Productiva Regional de la Guayaba y su Industria, Boyacá y Santander 2007 *Acuerdo Regional de Competitividad. Cadena Productiva de la Guayaba y su Industria. Boyacá y Santander* (Barbosa, Santander. Colombia)

[4] Gómez J C 2008 *Análisis de Necesidades Regionales en materia de Fluidos y Energía* (Bucaramanga, Santander: Colciencias/CDT de Gas)

[5] Fisher D 2002 *Commercial Cooking Appliance Technology Assessment* (EEUU: Food Service Technology Assessment FSTC)

[6] Manrique V 2011 *Diseño y Construcción de una Marmita Autogeneradora de Vapor, como alternativa para la Reconversión Tecnológica de la Industria del Bocadillo Veleño en Santander* Proyecto de Grado (Bucaramanga, Santander: UIS-Ingeniería Mecánica)

[7] F26 Committee 2012 ASTM F1602 - 12 *Specification for Kettles, Steam-Jacketed, 20 to 200 gal (75.7 to 757 L), Floor or Wall Mounted, Direct Steam, Gas and Electric Heated*

[8] American Society of Mechanical Engineers, ASME Boiler and Pressure Vessel Committee and Subcommittee on Pressure Vessels 2010 *Rules for construction of pressure vessels. International code VIII, Division 1 VIII, Division 1* (New York, N.Y.: American Society of Mechanical Engineers)

[9] ICONTEC 1997 *NTC 4082 - Equipos de Cocción para Uso Comercial que Funcionan con Gas. Requisitos de Seguridad* (Instituto Colombiano de Normas Técnicas y Certificación)

[10] F26 Committee 2008 ASTM F1785 - 97(2008) *Standard Test Method for Performance of Steam Kettles* (ASTM International)

[11] ICONTEC 2005 *NTC-ISO/IEC 17025. Requisitos Generales para la Competencia de los Laboratorios de Ensayo y Calibración.* (Instituto Colombiano de Normas Técnicas y Certificación)

[12] ONAC. Laboratorios de Calibración Acreditados. Certificado de Acreditación 10-LAB-013 Fecha de Actualización: 2010-08-19. Fecha de Vencimiento: 2015-08-18. (Organismo Nacional de Acreditación de Colombia).

[13] D03 Committee 2011 *ASTM D3588 - 98(2011) Practice for Calculating Heat Value, Compressibility Factor, and Relative Density of Gaseous Fuels* (ASTM International)

[14] D03 Committee 2010 *ASTM D1945 - 03(2010) Test Method for Analysis of Natural Gas by Gas Chromatography* (ASTM International)

[15] BIPM 2008 JCGM 100:2008 *Evaluation of measurement data — Guide to the expression of uncertainty in measurement*

[16] García L E 2009 Mediciones para la evaluación de la eficiencia de combustión de equipos generadores de vapor *METFLU Cienc. - Tecnol. - Innov.* **1**

[17] GRI Gas Research Institute 1997 *Method GRI-96/0008 - Determination of Nitrogen Oxides, Carbon Monoxide, and Oxygen Emissions from Natural Gas-Fired Engines, Boilers and Process Heaters Using Portable Analyzers*

[18] MRU Instruments, Inc. Mobile Fluegas Analysis. Handheld Analyzers – AMPRO 2000 Technical data (Houston, TX.)

IOP Conf. Series: Materials Science and Engineering **59** (2014) 012010 doi:10.1088/1757-899X/59/1/012010

Comparison of simplified models in the prediction of two phase flow in pipelines.

M. Jerez-Carrizales[1]*, J. E. Jaramillo[2] , D. Fuentes[3]
[1-2-3] Escuela de Ingeniería Mecánica, Universidad Industrial de Santander, Carrera 27 Calle 9, Bucaramanga, Santander, Colombia

E-mail: manuel.jerez@correo.uis.edu.co

Abstract. Prediction of two phase flow in pipelines is a common task in engineering. It is a complex phenomenon and many models have been developed to find an approximate solution to the problem. Some old models, such as the Hagedorn & Brown (HB) model, have been highlighted by many authors to give very good performance. Furthermore, many modifications have been applied to this method to improve its predictions. In this work two simplified models which are based on empiricism (HB and Mukherjee and Brill, MB) are considered. One mechanistic model which is based on the physics of the phenomenon (AN) and it still needs some correlations called closure relations is also used. Moreover, a drift flux model defined in steady state that is flow pattern dependent (HK model) is implemented. The implementation of these methods was tested using published data in the scientific literature for vertical upward flows. Furthermore, a comparison of the predictive performance of the four models is done against a well from Campo Escuela Colorado. Difference among four models is smaller than difference with experimental data from the well in Campo Escuela Colorado.

1. Introduction

The two phase flow is a complex phenomena. Interactions between two fluids generates structures commonly known as flow patterns. These flow patterns made the two phase flow more difficult to model than the single phase flow [1, 2].

Wells in Campo Escuela Colorado (Santander, Colombia) suffer from wax and asphalten deposition in the pipelines. For this reason, a project to deal with this problem was started. The main objective of that project is to develop some computational tools. It is expected that software predict material deposition rate in a section of the pipe. The prediction of the wax and asphalten deposition is going to be developed by two steps as shown in the figure 1. First, a model to predict conditions in every point of the pipe, second, a model to predict deposition at certain conditions of composition, pressure, temperature and velocity, besides others.

[1] *To whom any correspondence should be addressed.

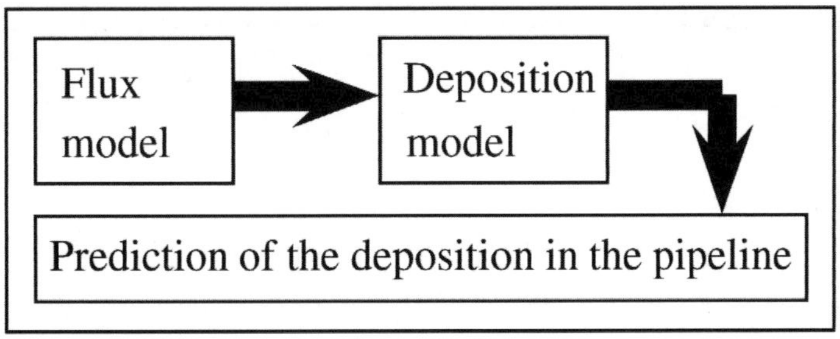

Figure 1. Steps in the project.

This work is focused in the first part. In order to check prediction of the pressure drop in a pipe a comparison between four models is done with a point of Campo Escuela Colorado. The goal in the two phase flow models in pipelines is to determine the pressure, temperature and volumetric flow profiles along it. In the oil industry, the total pressure drop allows a linkage between models for the reservoir and the pressure in the well head. Also, the volumetric flux is necessary to know the amount of liquid and gas produced by the well.

2. Models

The first model to predict two phase flow in vertical wellbores was developed by Poettmann & Carpenter [3] in 1952. Since then, several models have been published in the scientific literature. However, any model is not able to predict accurately the pressure drop for all conditions in the oil industry. In this work, three generally used models are selected and a new one too.

The equation to predict pressure drop in the pipe is shown in (1). The total pressure drop is the sum of the pressure drop due to gravity, friction and acceleration.

$$(dp/dz)_T = (dp/dz)_F + (dp/dz)_H + (dp/dz)_A \tag{1}$$

2.1. Hagedorn & Brown (HB) model

This model was published in 1965 and in spite of time elapsed, it was recommended by Ansari[4] in 1994. The HB model is flow pattern independent.

The original HB model [5] uses three graphs to calculate the viscosity number coefficient N_{LC}, the secondary correction factor ψ and the liquid holdup h_l (The holdup is determined as $h_l = \psi f(N_{LC},...)$). The holdup is used to determine a mixture density, a gravity pressure drop, a frictional pressure drop and an accelerating pressure drop.

Two modifications to the original method are here implemented:

- Since the method was coded in C++, the regression proposed by Guo et al [6] was used to replace the three graphs.
- The studied configurations are for vertical upward flow; therefore, the holdup must be greater than the non slip holdup.

2.2. Mukherjee & Brill (MB) model

The Mukherjee & Brill method [7] is a simplified model that is flow pattern dependent, and it can be used in deviated wells. The MB model is partly based on the adimensional numbers of Duns & Ros [8]. Furthermore, it defines new ones to predict transitions in the flow pattern as shown in the figure 2.

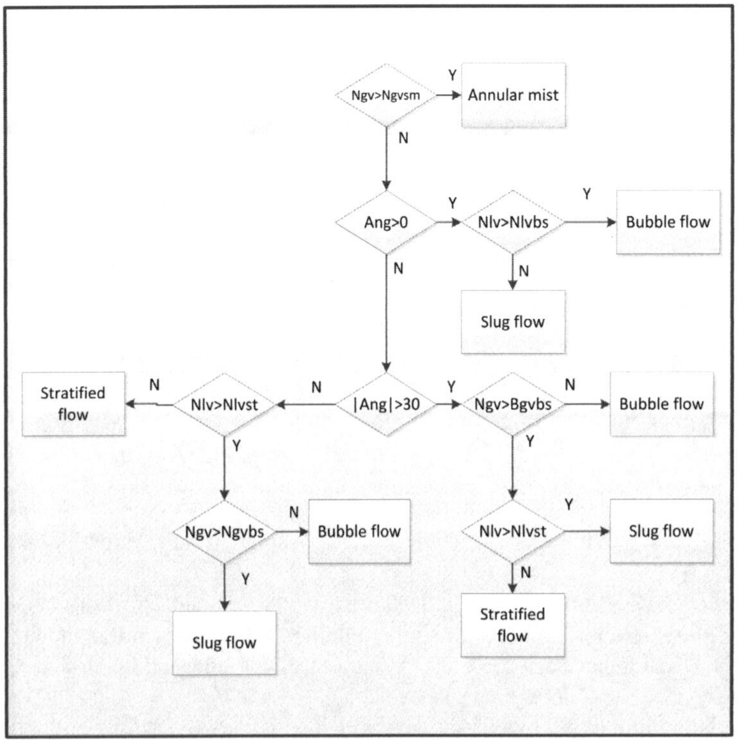

Figure 2. Algorithm to select flow pattern from MB model.

Liquid velocity number N_{lv}
Gas velocity number N_{gv}
Pipe diameter number N_d
Liquid viscosity number N_l
Liquid velocity number transition from bubble to slug flow $N_{lvB/S}$
Gas velocity number transition from slug to mist flow $N_{gvS/M}$
Transition from slug to stratified downflow $N_{lvS/T}$

This method uses an exponential equation to get the holdup. This equation depends on a group of coefficients; coefficients are selected whether flux is upward, stratified downward or other downward.

2.3. Ansari(AN) et al model
The Ansari et al [4, 1994] model is a mechanistic model based on the flow pattern map of Taitel et al and Barnea et al [9, 10]. The AN model shows a force balance for every flow pattern used.

The AN model includes many iterative subroutines, so, it demands more operations than the HB, MB and HK models to get the holdup and the pressure gradient.

2.4. Hasan et al (HK) model
The Hasan et al[11] model is based on the drift flux approach (2) for steady state conditions.

$$u_G = C_0\, u_M + u_D \tag{2}$$

The values for u_D and C_0 for (2) are selected from the table 1. They proposed an interpolation method for these values near a flow pattern transition.

Table 1. Values from the drift flux parameters by Hasan et al.

Flow pattern	C_0			u_D
	Upward	Countercurrent	Downward	
Bubble	1.2	2	1.2	$V_{\infty b}$
Slug	1.15	1.15	1.12	V_{∞}
Semianular	1.15	1.15	1.12	V_{∞}
Anular	1	1	1	0

3. Methodology

In this work, the models are implemented using the C++ object oriented programming philosophy. Correct models implementation should be validated. Therefore, models are compared with experimental data available in the literature [12, 13]. The general procedure is shown in the figure 3.

It is necessary to calculate physical, thermal, and chemical properties, flow pattern, holdup and pressure gradient at different points in the pipe for every data, as shown in figure 4. The validation step shows that results are influenced by the method applied to estimate fluid properties. Three correlations were analyzed, Standing, Al Marhoun and Vazquez & Beggs [2]. The properties that were calculated with these correlations were gas solubility and oil formation volume factor.

4. Results

4.1. Validation

4.1.1. Influence of properties. The table 2 shows the influence of the selected correlation applied to calculate fluid properties on the pressure drop. It is noted that mean error value is small; however, there are some points with a high error.

Table 2. Influence of methods to calculate properties

Method	Value [-]	HK [%]	MB [%]	HB [%]	AN [%]
Al Marhoun-Standing	Mean	3.64	3.45	2.52	3.03
	σ	4.60	3.44	4.84	4.46
	Max	17.26	14.47	16.31	17.17
Vazquez-Standing	Mean	1.76	1.86	3.25	1.92
	σ	2.23	2.61	5.49	2.43
	Max	8.12	10.37	21.25	9.76
Al Marhoun - Vazquez	Mean	3.86	3.88	2.60	3.82
	σ	4.73	4.47	3.92	4.62
	Max	15.41	15.92	15.22	15.84

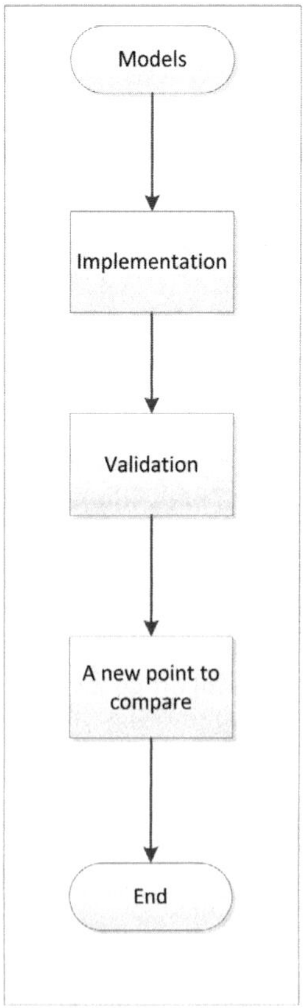

Figure 3. General procedure

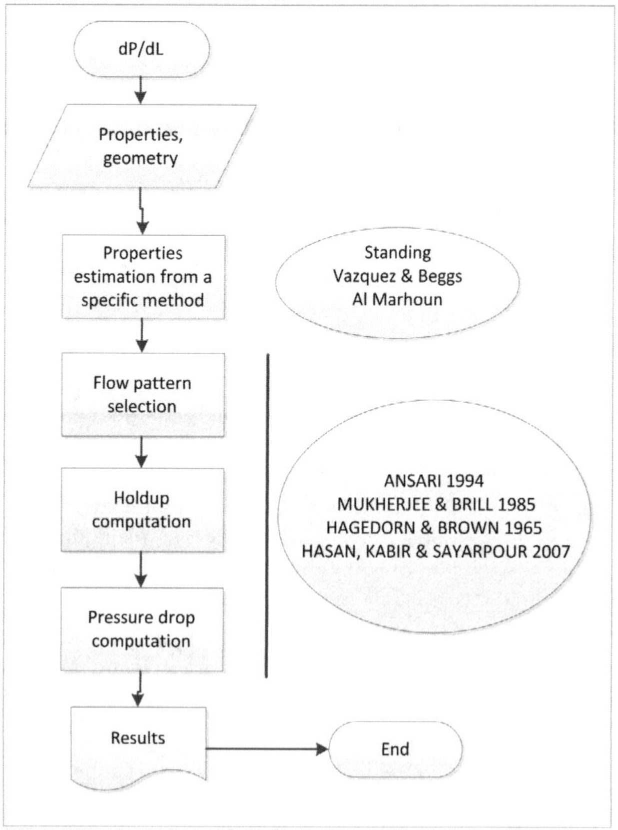

Figure 4. Procedure for every test

4.1.2. Pressure drop calculation. Four models with the same fluid properties were compared against experimental data to check implementation of the models. Results are shown in figure 5. The nearest to the central line the best the results.

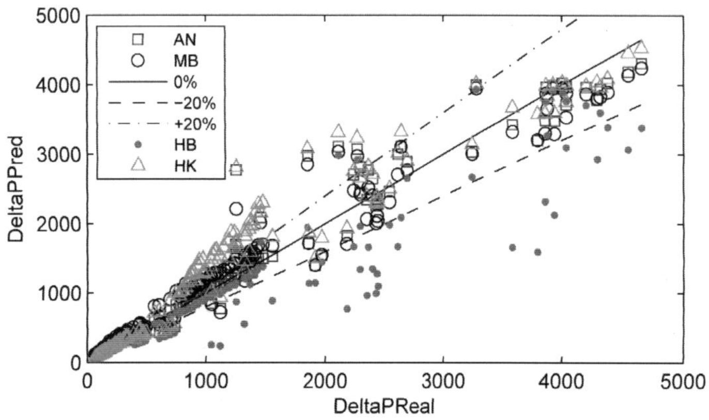

Figure 5. Comparison of measured pressure drop against values predicted by the Ansari et al, Mukherjee & Brill, Hagedorn & Brown and Hasan et al models.

Finally, a regression of data in figure 5 to test numerically the best model is made. This is shown in table 3. The best model has an *m* value and a determination coefficient R^2 near to 1.

Table 3. Linear regression of data from figure 5.

Model	m [-]	b [psi]	R^2[-]	σ_{error} [psi]
AN	0.9713	53.1976	0.9521	266
MB	0.9076	186.3821	0.9538	244
HB	0.7594	42.6434	0.8584	377
HK	0.9898	185.3421	0.9306	330

The best models are the Ansari et al model and the Hasan et al model.

4.2. Comparison with Well 37 from Campo Escuela Colorado

Experimental data selected for comparison purposes are those from [14]. The values are measured in the well number 37 at Campo Escuela Colorado. Two different tests were done by the ICP (Instituto Colombiano del Petróleo) in January 30 of 2008, a mechanical state and a PVT (pressure-volume-temperature) analysis. Data measured were pressure and temperature from wellhead, reservoir and bottomhole. Furthermore, oil flow, gas flow, water flow, API gravity, gas components, tubing diameter and length of the pipe were determined.

In figure 6 variation of predicted pressure for each model with depth is presented together with experimental data at the bottom of the well 37 (grey diamond). As it can be seen all models over-predict results. The difference between the predicted and measured pressure at the well bottom is around 30%, and all models predict a similar value: $\%error_{AN}=33$, $\%error_{MB}=30$, $\%error_{HB}=30$, $\%error_{HK}=33$. Moreover, holdup predicted by the four models is presented in figure 7. This graph shows throughout the well two phase flow is present due to holdup never reach the value of one that indicates a single phase is achieved.

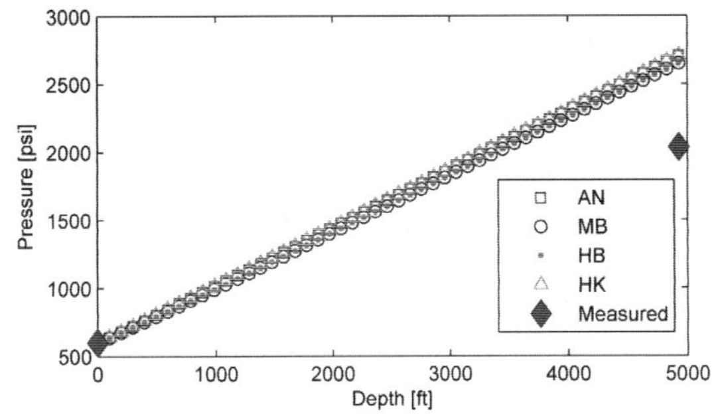

Figure 6. Pressure predicted by the AN, MB, HB, HK models.

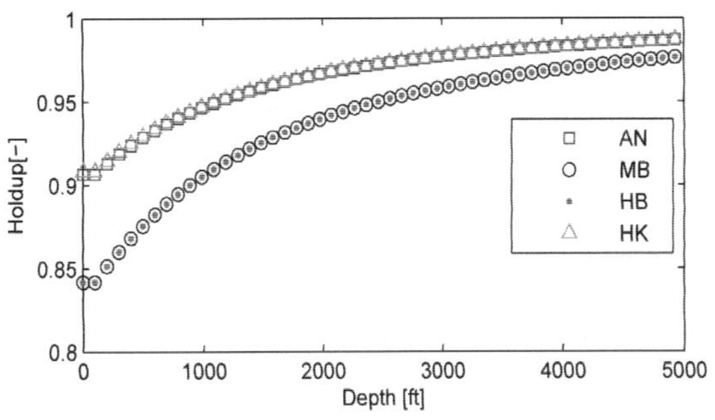

Figure 7. Holdup predicted by the AN, MB, HB, HK models

5. Conclusions

1. It has been observed that fluid properties (correlations used) has an influence in the total pressure drop that was a 3% in average, but that can be as high as 20 %.
2. The best performance was presented by the Ansari et al and the Hasan et al model in the validation process. Furthermore, the first one is simpler than the second one method, and it does not need internal iterative routines.
3. When the models were compared with data from the well 37 of Campo Escuela Colorado all models exhibited a difference of about 30%. Therefore, it should be made a deeper analysis to better understand the source of these differences for this case.

6. References

[1] Ishii M and Hibiki T 2011 *Thermo-fluid dynamics of two-phase flow* (Springer New York) ISBN 978-1-4419-7984-1

[2] Mukherjee H and Brill J P 1999 *Multiphase Flow in Wells* (Henry L. Doherty Memorial Fund of AIME)

[3] Poettmann, Fred H;Carpenter, Paul G 1952 *Drilling and Production Practice* (American Petroleum Institute) pp 257–317 URL http://www.onepetro.org/mslib/servlet/onepetropreview?id=API-52-257

[4] Ansari, AM ; Sylvester, ND ;Sarica, C, Shoham, O , Brill, JP 1994 *SPE Production & Facilities* **Volume 9**, **Number 2** p. 143–151

[5] Hagedorn, Alton R; Brown, Kermit E 1965 *Journal of Petroleum Technology* **Volume 17**, **Number** 4475–484

[6] Guo, Boyun; Lyons, William C; Ghalambor, Ali 2007 *Petroleum Production Engineering* (Elsevier Science & Technology Books) ISBN 978-0-7506-8270-1

[7] Mukherjee H and Brill J P 1985 *Journal of energy resources technology* **107** 549–554 ISSN 0195-0738

[8] Duns, H J; Ros, N 1963 *6th World Petroleum Congress* (World Petroleum Congress) pp p. 451–465

[9] Barnea D, Shoham O and Taitel Y 1982 *Chemical Engineering Science* **37** 741 – 744 ISSN 0009-2509 URL http://www.sciencedirect.com/science/article/pii/0009250982850343

[10] Taitel Y, Bornea D and Dukler A 1980 *AIChE Journal* 26 345–354 URL http://onlinelibrary.wiley.com/doi/10.1002/aic.690260304/abstract

[11] Hasan, AR; Kabir, CS; Sayarpour, M 2007 *SPE Annual Technical Conference and Exhibition* (Society of Petroleum Engineers)

[12] Baxendell P B; Thomas R 1961 *PETROLEUM TRANSACTIONS* p. 1023–1028

[13] Aziz, Khalid; Govier, G W; Fogarasi, M 1972 *Journal of Canadian Petroleum Technology* **Volume 11, Number 3**

[14] Oviedo G and Chaparro C 2011 *Modelado y simulación del flujo de hidrocarburos en tuberías mediante el método de Mukherjee & Brill.* (Universidad Industrial de Santander, Escuela de Ingeniería Mecánica, tesis pregrado)

Acknowledgments

This work received financial support provided by VIE-CEC Vicerrectoría de Investigación y Extensión – Campo Escuela Colorado No. 8557, Universidad Industrial de Santander, and, Jóvenes Investigadores Colciencias No. 617.

IOP Conf. Series: Materials Science and Engineering **59** (2014) 012011 doi:10.1088/1757-899X/59/1/012011

Quality evaluation of energy consumed in flow regulation method by speed variation in centrifugal pumps.

S Morales[1]*, M Culman[2], C Acevedo[3], C Rey[4]

[1-2-3-4]Research Group in Resources, Energy and Sustainability (GIRES), Faculty of Energy Engineering, Universidad Autónoma de Bucaramanga (UNAB), Bucaramanga, PC: 680003, Colombia.

E-mail: smorales@unab.edu.co

Abstract. Nowadays, energy efficiency and the Electric Power Quality are two inseparable issues in the evaluation of three-phase induction motors, framed within the program of Rational and Efficient Use of Energy (RUE).The use of efficient energy saving devices has been increasing significantly in RUE programs, for example the use of variable frequency drives (VFD) in pumping systems.The overall objective of the project was to evaluate the impact on power quality and energy efficiency in a centrifugal pump driven by an induction three-phase motor, using the flow control method of speed variation by VFD. The fundamental purpose was to test the opinions continuously heard about the use of flow control methods in centrifugal pumps, analyzing the advantages and disadvantages that have been formulated deliberately in order to offer support to the industry in taking correct decisions. The VFD changes the speed of the motor-pump system increasing efficiency compared to the classical methods of regulation. However, the VFD originates conditions that degrade the quality of the electric power supplied to the system and therefore its efficiency, due to the nonlinearity and presence of harmonic currents. It was possible to analyze the power quality, ensuring that the information that comes to the industry is generally biased.

1. Introduction

As development of electronics proceeds electronics and power generation is becoming increasingly expensive, it is necessary to increase efficiency in production processes. This problem is particularly important in the case of applications where pumps are used, since this would allow a reduction in power consumption and more efficiency in the motor and pumping system. The traditional methods used for flow control in pumps are based on the modification of the characteristic curve of the system: Pressure (H) v / s flow rate (Q), which is achieved by using one or more valves, for example, the method of regulating by pressure throttling and bypass flow control method. The use of these methods results in an increase in power consumption, which leads to a significant reduction in efficiency of the motor-pump group [1].

A more efficient way to regulate pump flow is by changing the speed of the electric motor driving the pump, using a three-phase squirrel-cage induction motor, along with a VFD.

Currently, one of the trends in the industry is precisely the use of VFD in pumping systems, as they pertain to special devices lists in Energy Efficiency programs [1-4]. Its principal advantage over the other methods of regulation is that it does not require an intervention in the hydraulic system via

[1] *To whom any correspondence should be addressed.

valves and other electro-mechanical elements. When the device is connected to the electric grid, you can operate the system and achieve the optimal operating point (the most efficient) [5]. However, according to the hydraulic characteristics of the system, its operation can be uneconomical, because VFD are characterized by altering the electric power quality, due to the high magnitudes of harmonic currents that appear in grid side , as a result of the strong distortion suffered by the consumed current [6]. It has been demonstrated that generate harmonics of order 5 and 7 to 63% and 33 % of the fundamental component of the current, respectively, which puts into question the true energy savings provided by the use of VFD in flow regulation in pumps [1, 2, 7]

Additionally, companies that require the application of a flow control method can make wrong decisions because of the technical information provided by the commercial interests of the companies that manufacture VFD, or ignore the requirements or additional costs that must be taken into account in the purchase and implementation of a VFD [1, 2, 4]. In this vein, there is concern of how advisable is the practice of flow control method by speed variation with a VFD from the point of view of its impact on the quality of electric power consumed by the grid.

2. VFD in pumping systems

A VFD system (also called ''frequency converter'') is a device which controls the speed of an AC motor by regulating its voltage input and frequency. This allows the electric motor to reach the optimal operating point (maximum efficiency) and reduce operation costs thanks to energy savings. These devices provide a high degree of control over a pumping system by easily adapting to flow and pressure fluctuations, which is a condition that traditional mechanical methods such as throttling and bypass cannot offer. This causes manufacturers of VFD to offer them as a practical and efficient solution for flow regulation in many industrial applications. However, VFD are not advisable for all applications, for example systems that operate high static head and those that operate for long periods under low flow conditions [8, 9].

Also, as it was stated before, VFD are known to produce conditions that deteriorate the quality of the energy consumed due to non-linearity the presence of high levels of harmonic distortion. In three-phase VFD there are harmonic distortion of 5, 7, 11, 13, 19 and higher odd multiples of the fundamental frequency (60Hz) [8].

Harmonic distortion can occur in voltage or current sine waves. In this case, the focus is on distortion in current wave, which is injected to the feed-in network by the users. These harmonic currents flow into the electric network and local electric supply lines and produce negative effects such as overheating of electric conductors, voltage drops, overheating, low efficiency and premature failure of motors and transformers, interference in measurement and communication equipment, etc.

The use of VFD in pumping systems can help to save energy and money because it is an efficient way to regulate flow according to the specific conditions of a process, but the deterioration of the energy consumed can question the its advantages and performance, especially in meeting energy quality standards.

3. Description of the system

The tests were performed on a small scale hydroelectric generation unit composed of Francis and Kaplan turbines. The main components of the system are described below.

3.1. Hydraulic generation and pumping system

The system in which the tests were performed consists of a hydraulic bench containing a unit of small scale Francis and Kaplan turbines for a laboratory. In this case, it was determined that the system would be modeled using the Kaplan turbine which is coupled to an electric generator. Thus, the curve of the system is limited by the pressure drop, due to the transformation that the Kaplan turbine makes of fluid energy into mechanical energy [1, 2]. The scheme of the system operating with the Kaplan turbine is shown in figure 1.

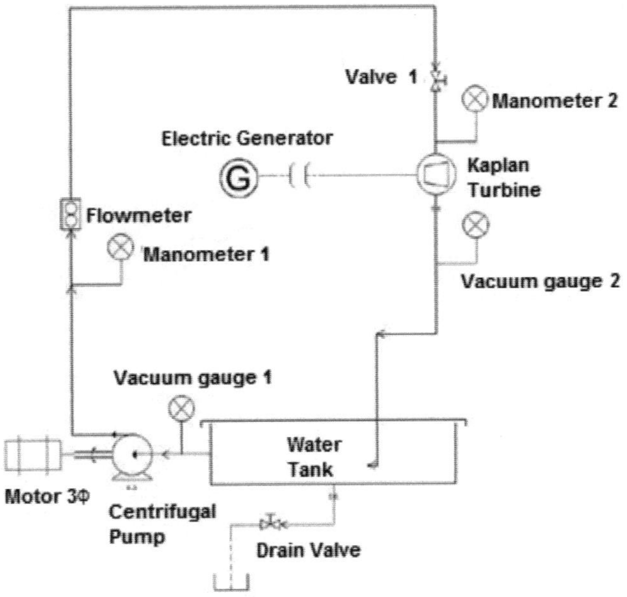

Figure 1. Scheme of the hydraulic generation system operating with Kaplan Turbine [1, 2]

As seen in figure 1, in the suction and the discharge lines there are pressure gauges are available for measuring static pressure or vacuum. Additionally, there is a flowmeter at the pump outlet indicating current flow [1, 2]

3.2 The electric motor

The three-phase induction motor that drives the pump has the following technical characteristics:

- Four poles, Three-phase
- Voltage:220Δ/380Y V.
- Current: 68.8Δ /39.8Y A
- Power: 20HP-15KW
- Power factor: Cos □ 0.8
- Frequency: 60Hz
- Nominal speed: 1750 rpm
- Insulation class F and protection IP 54
- Design class B according to NEMA classification.

3.3 The centrifugal pump

The centrifugal pump used for the tests has a nominal speed of 1750 rpm. The table 1 shows the nominal characteristics.

Table 1. Pump nominal information [1, 2, 10]

H (m)	Q(m³/h)
Hmax 31,5	Qmin 48
Hmin 14	Qmax 210 m

The pump was manufactured according to the European Standard EN 733 aspiration centrifugal electropumps [1, 10, 11]. The figure 2 shows the pump's characteristic curve supplied by the manufacturer. In this case, the type B pump was used, with a discharge diameter of 267mm.

The figure 3 shows the performace of the pump at a speed of 1750 rpm for the type B pump.

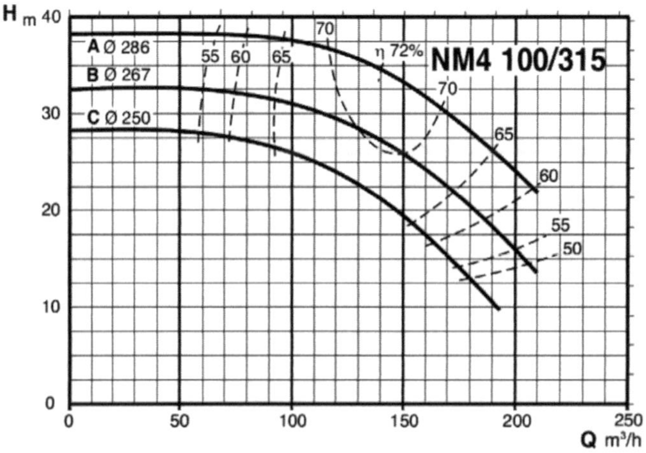

Figure 2. Head vs Flow and Efficiency [1, 2, 10]

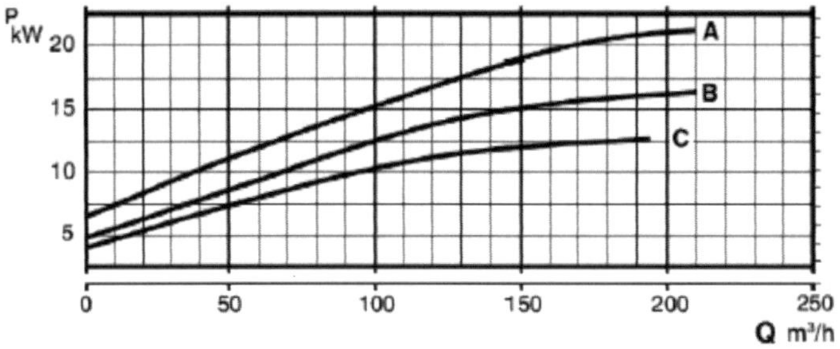

Figure 3. Power vs Flow [1, 2, 10]

4. Methodology

For this case, the operation of the motor-pump group was made with the VFD. The measurement of the electric power quality was made taking into account only the emission levels of harmonic currents using a three-phase power quality analyzer. This procedure is regulated according to IEC 61000-3-4, which sets limits for harmonic currents on low and medium voltage equipment consuming more than 16 A per phase and includes all equipment connected to the public network low-voltage alternating current, rated voltage up to 240 V rated voltage single-phase and three-phase to 600 V 50 Hz or 60 Hz [11].

During this test, valve 1 was kept in a fully open position and the tank of the system was kept full to ensure continuous supply of water. In order to obtain enough operating points of the pump, the VFD was programmed to change the rotation speed according to a range of frequencies between 0 Hz and 60 Hz, through a linear potentiometer and consistent monitoring of the frequency output, which corresponds to 21 points of operation in a flow range from $30m^3/h$ to $127.5\ m^3/\ h$.

For each operating point, the following data was registered [2]:

- Motor speed in revolutions per minute.
- Active, reactive and apparent power and power factor before the VFD.
- Voltages and currents before the VFD.
- Frequency and peak factor.
- Odd harmonics of voltage and current.
- Voltage, current and power output percentage of the drive.
- Output frequency of the inverter.

Data were collected before and after the VFD. Because this device is part of the energy balance, it was necessary to record the parameters that it registers in the output. Considering that there is a derating factor for the VFD, similar to the K factor transformers, which indicates how much to reduce the maximum output power when there are harmonics. It is understood that the useful power received by the motor is the measured active power (KW) and not calculated through the three-phase power formula $(\sqrt{3})\ V_L I_L$ [1, 2].

According to IEC 61000-3-4 standard, the evaluation method of harmonic currents can be performed with direct measurement of the emission or calculating the emission by simulation [13]. If direct measurement is made, it must be done in normal operation conditions for a week, where 95% of the RMS values of each harmonic voltage averaged in 10 minutes should not exceed the indicated values. However, for this test, the measurement period was limited to the time needed to take the information from different points of flow.

The connection between the device under test and its power supply is direct. But in this case, the electric feed-in of the hydraulic generation system comes from a circuit that has other loads of the building connected in parallel. Therefore, it cannot be ensured that there is immediate connection to the point of common coupling and isolated from other loads. The system was connected to a voltage of 220V and a frequency of 60 Hz [2].

In Colombia, in the mid-nineties the generating and distribution companies addressed the subject of the Power Quality, in view of technological developments through microelectronics that arrived to the country. The Energy and Gas Regulatory Commission (CREG), adopted the guidance of the Institute of Electrical and Electronics Engineers - IEEE 519-1992 about harmonic limits, through its Resolution 070 of 1998 [11]. Therefore, the harmonic current limits on which this article is based are set in IEEE 519 and IEC 61000-3-4 standards. Tables 2 and 3 present the harmonic current limits established in the standards IEE5519-1992 and IEC 61000-3-4 and respectively.

Table 2. Maximum harmonic distortion of current in percentage according to IEEE 519 [2, 4, 13, 14].

	Odd harmonics of orders 3rd to 11th	**Total Harmonic Distortion in Current (THDi)**
Limit of current distortion for general Distribution systems (120V to 69000V).	4,0	5,0
Limit of current distortion for general sub-transmission systems (69001V to 161000V).	2,0	2,5
Limit of current distortion for general transmission systems (69001V to 161000V), dispersed generation and cogeneration.	2,0	2,5

According to table 2, the maximum limit of DTH (%) taken as reference in this case is 5%, for general distribution systems 120V to 69000V. For limits of individual harmonic currents, the values of reference in this case are set out in table 3.

Table 3. Maximum distortion of individual harmonics according to IEC 61000-3-4 [2, 13, 15]

Harmonics non multiple of 3		**Harmonics multiple of 3**	
Harmonic order (n)	**Maximum current In/I_1 (%)**	**Harmonic order (n)**	**Maximum current In/I_1 (%)**
5	10,7	3	21,6
7	7,2	9	3,8
11	3,1	15	0,7
13	2,0	21	≤ 0,6
17	1,2	27	≤ 0,6
19	1,1	≥ 33	≤ 0,6
23	0,9		
25	0,8		
29	0,7		

5. Results and analysis

Table 4 shows the results obtained during the test for 21 operation points regarding the power quality in terms of harmonic distortion. The harmonics correspond to the average obtained from the three line currents.

As seen in Table 4, the levels of total and individual harmonic distortion decrease as the frequency is about to reach to 60 Hz, which is the nominal frequency of operation of the equipment and the power grid.

Table 4. Results of current harmonics due to the VFD [2]

Operation Point	Q (m^3/h)	F (Hz)	Total Harmonic Distortion and Individual Odd Harmonics THD(%)	H3(%)	H5(%)	H7(%)	H9(%)	H11(%)	H13(%)	H15(%)
1	30	16,3	175,93	23,03	93,23	87,03	17,93	71,80	62,37	10,80
2	35	18,4	169,27	18,60	92,40	85,33	14,13	67,93	57,63	8,30
3	40	20,5	162,30	23,03	91,73	84,27	16,87	66,00	55,30	8,93
4	45	22,7	147,20	14,77	90,27	81,27	10,30	59,27	40,50	5,37
5	50	25,1	146,33	15,17	89,57	80,13	10,37	38,90	44,97	5,23
6	55	27,2	137,13	13,43	87,90	77,07	8,43	51,30	38,27	4,03
7	60	29,3	136,47	16,23	87,67	76,67	9,87	50,77	37,67	4,40
8	65	31,6	127,50	13,10	85,63	73,07	7,77	44,53	31,07	3,57
9	70	33,1	125,53	12,73	85,20	72,17	7,13	42,83	29,27	3,30
10	75	35,7	115,40	12,63	82,83	68,20	6,10	36,53	23,13	3,00
11	80	38,1	112,80	10,43	80,87	64,57	5,33	31,20	18,27	2,80
12	85	40,3	107,93	8,83	79,63	62,70	4,10	28,73	16,20	2,17
13	90	42,4	102,27	13,50	77,23	58,43	5,50	23,10	11,93	3,37
14	95	44,8	98,70	14,17	75,50	55,67	4,87	19,53	9,37	3,03
15	100	46,6	95,50	13,33	74,17	53,67	4,33	17,37	8,23	2,73
16	105	48,8	91,23	12,50	72,40	50,93	3,93	14,60	7,10	2,37
17	110	50,9	90,50	4,67	72,50	50,93	1,80	14,10	5,63	0,90
18	115	52,9	89,43	4,90	72,03	50,20	1,73	13,27	4,63	0,77
19	120	55,5	85,60	4,23	69,93	47,07	1,40	10,33	3,97	0,63
20	125	58,1	82,60	3,83	68,53	45,07	1,13	8,70	3,73	0,43
21	127,5	60	80,80	3,77	67,53	43,63	1,43	7,63	3,67	0,53

Figure 4 shows a comparison of the measurement results of THD and individual odd currents with the maximum limits established in IEEE 519 and IEC 61000-3-4 standards.

According to the results presented in figure 4, it shows that VFD devices inject large amounts of harmonics in the electrical network. The values of THD express the large amount of harmonic current flowing in the network, for example, the operating point of 30 m^3/ h has a THD of 180%, which means that the harmonic currents are almost two times higher than the fundamental current [2]. This also reflects not only the size of disturbances in the grid, but also that the VFD is characterized for having large energy losses represented in the power dissipated by harmonic currents that do not contribute to the useful power.

Also, a tendency to lower the current harmonics as the flow increases was identified. This means that the quality of the power improves as it approaches the nominal operating conditions, at a frequency of 60 Hz. Likewise, the values of the harmonic multiples of 3 are lower than the amounts registered in the other orders. The high levels of harmonics multiples of 3 indicate that the electric motor is suffering pro and against torque forces, which causes it to lose power and overheat when it is operated using the VFD [2].

IOP Conf. Series: Materials Science and Engineering **59** (2014) 012011 doi:10.1088/1757-899X/59/1/012011

The red dots in the figure 4 represent the harmonic emission limits established in IEEE 519 and IEC 61000-3-4 standards. It can be observed that THD limits established by the IEEE 519 Standard were not accomplished in any operation point.

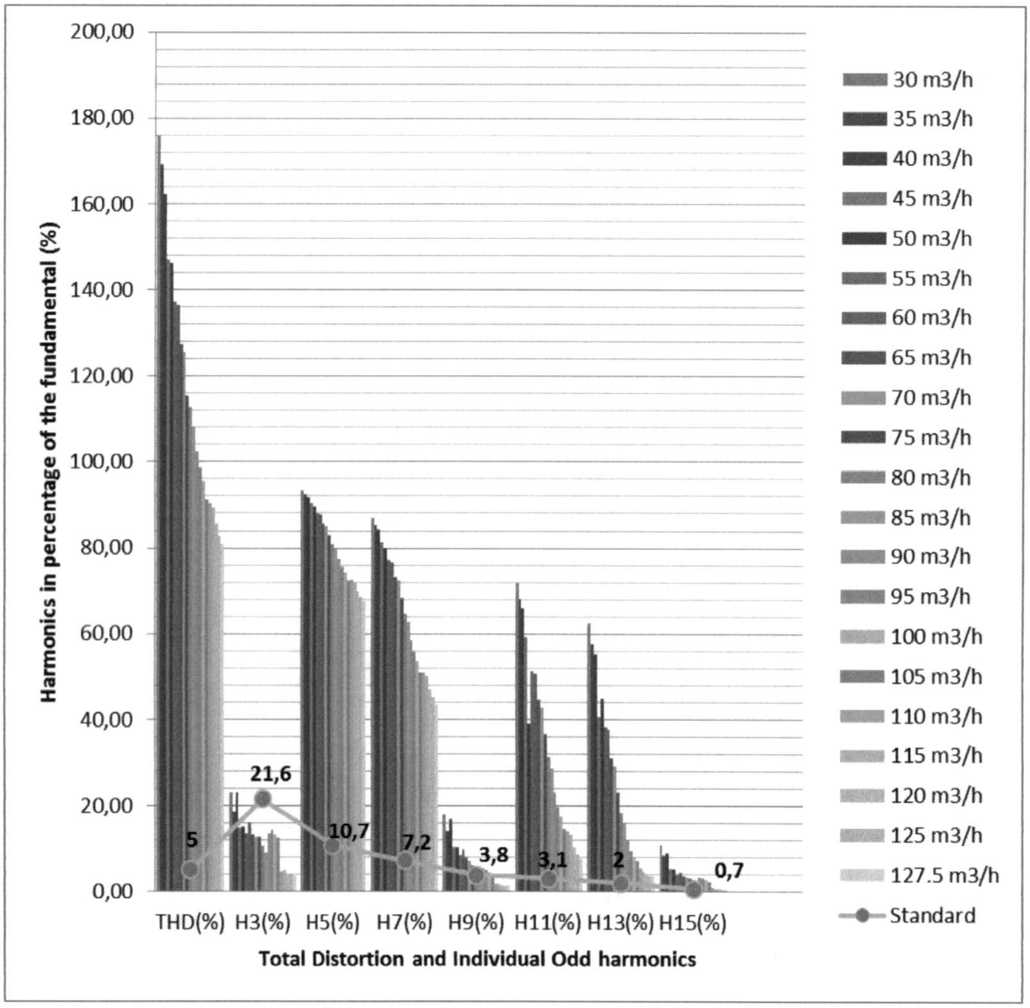

Figure 4. Comparison of THD and individual odd harmonic levels with IEEE 519 and IEC 61000-3-4 Standards [2]

The limits of individual harmonic currents of order 5-15 and at some points of order 3 were massively unaccomplished according to IEC 61000-3-4 Standard. As a particular case, it is recommended a prior diagnosis of the system's grid, which can be done by connecting a purely resistive load to identify and evaluate the harmonic currents that may be circulating and are unrelated to the pumping system [2].

6. Conclusions

With the development of this project, it was possible to put into question the effects that bring the use of VFD on the electric power quality. The use of VFD for flow regulation in pumping systems has as the main advantage, the energy savings, but the emission of large amounts of harmonics deteriorates its performance.

According to the information gathered, it was shown that VFD distorts the voltage sine wave and produces harmonic currents. In order to mitigate these problems, it is necessary to acquire additional devices such as filters for the elimination of harmonics or the use of shielded cables to conserve electromagnetic compatibility, which increases operation costs.

In terms of electromagnetic compatibility standards, it was found that the conditions established in the IEC 61000-3-4 standard in its execution to ensure its validity contains requirements that are very difficult to meet completely because of their stringency.

Finally, it was possible to identify that there is a need to consider additional costs that companies have to assume in order to ensure the quality of electric power when using a VFD. This must be done before purchasing and installing the VFD, since solving the problems caused by the presence of harmonics after having installed the machine usually consumes more time and money.

Acknowledgements

The authors express their acknowledgements to Universidad Autónoma de Bucaramanga (UNAB) and the Program "Joven Investigador" of COLCIENCIAS for supporting the activities which resulted on the writing of this article.

References

[1] Culman M, Morales S, Rey C and Acevedo C 2013 Evaluation of the impact of power quality and energy effienciency in flow regulation methods in pumps *ERMA* (in evaluation).

[2] Culman M 2013 *Evaluation of the impact of power quality and energy effienciency in flow regulation methods in pumps*. Bachelor Thesis (Universidad Autónoma de Bucaramanga, Colombia).

[3] Electroindustria *2008 Variadores de frecuencia: responsables de perfeccionar la acción de los motores*. Editorial Group EMB August 28th edition (Santiago, Chile)

[4] Hoevenaars T, LeDoux K and Colosino M 2003 *Interpreting IEEE Std 519 and Meeting its Harmonic Limits in VFD Applications*. Petroleum and Chemical Industry Conference. Record of Conference Papers. IEEE Industry Applications Society 50th Annual: Institute of Electrical and Electronics Engineers - IEEE 145-150.

[5] Mantilla L and Cardona J *Gestión Energética de los motores eléctricos: mejora de la eficiencia de los accionamientos con el uso de variadores de velocidad*. Department of Electric and Energy Enginneering. University of Cantabria. (Santander, Spain)

[6] Blanco E, Verlarde S and Fernández J 1994 Sistemas de bombeo. Department of Energy, University of Oviedo. (Gijón, Spain)

[7] Universidad Francisco José de Caldas. *Los armónicos en las instalaciones eléctricas*. Available: http://www.udistrital.edu.co:8080/documents/42573/87032/T025+LOS+ARMONICOS+EN+LAS+INSTALACIONES+ELECTRICAS (Consulted: 04-02-2013)

[8] Camoinaro R and Dellepiane G 2005 Variable frequency drives for MSF desalination plant and associated pumping stations. *Desal* **182** 53-65.

[9] U.S Department of Energy and the Hydraulic Institute 2006 *Improving pumping system performance: a sourcebook for industry*. Lawrence Berkeley National Laboratory, Resource Dynamics Corporation, Alliance to Save Energy and the Hydraulic Institute Washington D.C. U.S Department of Energy 25-35.

[10] Calpeda. *NM4 60 Hz Close Coupled Centrifugal Pumps*. Available: http://it.calpeda.com/file/aaecc9cb5988b132b7208ff162ac639a.pdf (Consulted: 19-07-2013)

[11] Flórez I 2011 *Análisis Comparativo de los Métodos de Medición "In Situ" de la Eficiencia en Motores Trifásicos de Inducción*. Bachelor Thesis. (Universidad Autónoma de Bucaramanga, Colombia)

[12] Torres H, et.al 2005 *Energía eléctrica, un producto con calidad -CEL-*. (ICONTEC: Colombia)

IOP Conf. Series: Materials Science and Engineering **59** (2014) 012011 doi:10.1088/1757-899X/59/1/012011

[13] Pomilio J. *Electromagnetic Compatibility (EMC) Low-Frequency Standards*. (Universidade Estadual de Campinas, Brazil) http://www.dsce.fee.unicamp.br/~antenor/pdffiles/IEC.pdf (Consulted: 6-08-2013)
[14] Blooming T, Carnovale D 2006 *Application of IEEE Standard 519, Harmonic Limits*. (U.S.A, Atlanta)
[15] Enríquez G 1999 *El ABC de la calidad de la energía eléctrica*. (Limusa: Mexico) 84-88
[16] IEC-International Electrotechnical Commission 1998 *International Standard, IEC 61000-3-4, Electromagnetic compatibility* (Geneva, Switzerland).

IOP Conf. Series: Materials Science and Engineering **59** (2014) 012012 doi:10.1088/1757-899X/59/1/012012

Analysis of energy production with different photovoltaic technologies in the Colombian geography.

Y Muñoz[1]*, D Zafra[2], V Acevedo[3] and A Ospino[4]

[1-2-3]Research Group in Resources, Energy and Sustainability (GIRES), Faculty of Energy Engineering, Universidad Autónoma de Bucaramanga (UNAB), Bucaramanga, PC: 680003, Colombia.
[4] Renewable Energy, Universidad de la Costa CUC, Barranquilla, Colombia.

E-mail: ymunoz294@unab.edu.co

Abstract. This research has analyzed the photovoltaic technologies, Polycrystalline silicon, Monocrystalline Silicon, GIS, Cadmium Tellurium and Amorphous Silicon; in eight cities of the Colombian territory, in order to obtain a clear idea of what is the most appropriate for each city or region studied. PVsyst simulation software has been used to study in detail each photovoltaic technology, for an installed capacity of 100kW knowing the specific data of losses by temperature, mismatch, efficiency, wiring, angle inclination of the arrangement, among others

1. Introduction

Colombia is a country with a great diversity of scenarios in both climate and energy from the solar perspective. The cities are located in places ranging from almost desert, (Riohacha - Guajira) up to plateaus of low temperature (Tunja - Boyaca). An analysis of photovoltaic technologies in the different cities of the Colombian territory can give a clear idea of which photovoltaic technology is most suitable for each city or region studied. Using PVsyst simulation software, it is possible to study in detail each Photovoltaic (PV) technology [1], knowing specifics of losses by temperature, mismatch, efficiency, wiring, angle inclination of the array, among others, by analyzing the energy performance and not the area according to efficiency[2].

The selected cities have different characteristics in different areas of the country, climates, and hydrological conditions [3]. The selected zones in this study are:

- Arauca – Arauca.
- Bahía Solano – Chocó.
- Barranquilla – Atlántico.
- Bogotá – Cundinamarca.
- Bucaramanga – Santander.
- Guajira.
- Leticia – Amazonas.
- Tunja – Boyacá

[1] *To whom any correspondence should be addressed.

This study shows that a photovoltaic system in a city with a low radiation can produce more energy per square meter than other with higher radiation. This means that there is an opportunity to apply this technology in different thermal conditions of the Colombian geography.

2. Description of the study

Characteristics and coordinates of places taken as the basis for this analysis.

Table 1. Meteorological data of studied cities.

CITY-REGION	COORDINATES	AMBIENT TEMPERATURE (C)	WIND VELOCITY (m/s)	ELEVATION (masl)	ANNUAL SOLAR IRRADIATION (kWh/year)	PSH (h)
ARAUCA	7.08N 70.75W	24.76	3.2	360	1762.5	4.83
CHOCO-BAHIA SOLANO	6.29N 77.47W	24.5	1.8	60	1530.5	4.19
BARRANQUILLA	10.98N 74.78W	27.4	4.4	50	1922.9	5.27
BOGOTA	4.23N 74.49W	15.3	3.4	2650	1769	4.85
BUCARAMANGA	7.13N 73.13W	23.3	1.7	965	1881.5	5.15
GUAJIRA	12.01N 71.7W	26.1	6.2	30	2370.5	6.49
LETICIA	4.21S 69.9W	27.7	1.4	90	1670.7	4.58
TUNJA	5.54N 73.36W	12.9	1.8	2780	1837.6	5.03

Table 1 shows valuable data when determining in which region a photovoltaic installation would work better. The study of La Guajira region presents the highest value at Peak Solar Hours, so it is expected that the energy production there would be higher than in the rest of the cities studied.

As starting points for PV evaluation, and leaving photovoltaic technology as the only variable, the following considerations have been taken [4]:

- Installed capacity of 100W
- Inverters for grid connection KACO POWADOR 36TL
- Constant losses: dirt, ohmic, miss match, heat.

PVsyst allows different variants of simulation. The analysis has been made with 2 variants of design: with and without the speed of wind.

For the study of the behavior of photovoltaic technologies in different cities, the main photovoltaic modules used in the market have been taken as object of study [5]. The photovoltaic modules used for this study are shown in table 2.

The characteristic curves for each photovoltaic module, analysis of the behavior of each technology with variables such as temperature and incident radiation are available in the PVsyst software [6].

IOP Conf. Series: Materials Science and Engineering **59** (2014) 012012 doi:10.1088/1757-899X/59/1/012012

Table 2. Photovoltaic modules used in the study

Panel	Fabricant	Technology	Power (w)	Panel area (m^2)	Temperature coefficient of panel (%/c)	Module efficiency
JAM6-60-250	JA SOLAR	Monocrystalline	250	1,635	-0,44	15,54%
JAP6-60-250	JA SOLAR	Polycrystalline	250	1,635	-0,45	15,31%
FS-390	FIRST SOLAR	Cadmium telluride (CdTe)	90	0,72	-0,25	12,52%
DA100-A1	DUPONT APOLLO	Amorphous silicon (a-Si)	100	1,564	-0,25	6,40%
SF165S	SOLAR FRONTIER	CIS	165	1,228	-0,31	13,44%

Two case of study are analyzed, the case A presents the conventional analysis of PV energy production without consider the wind speed, oppositely the case B shows the effects of consider the wind speed.

3. Case study A: 100kW without influence of wind speed

3.1. Energy Production

The figure 1 shows the amount of energy produced by the arrays in case study A. In terms of the amount of power generated, the department of La Guajira has higher energy production in all technologies. On the other hand, the smallest photovoltaic energy production of the studied areas is Bahía Solano in the department of Chocó, corresponding to the data of lower incident radiation and high temperatures in the region.

As for Bogota and Arauca, they do not have equal energy production results due to the difference of temperature between these regions, that are completely opposite, being the low temperatures of Bogota more favorable for the performance (produces 4% more in monocrystalline technology and a 1.7% more in CdTe technology) and behavior of photovoltaic modules regardless of the type of technology used.

When comparing Bucaramanga and Tunja, it can be said that despite Bucaramanga has higher radiation, Tunja presents greater energy production on behalf of the temperature. There is a great difference between monocrystalline production (in Tunja produces 3.8% more energy than in Bucaramanga) and thin-film technologies, especially type Telluride CdTe (0.69% more than in Bucaramanga), making this last technology more suitable for warm climates.

Performance ratio (PR) is defined the relation between the energy actually produced and the energy that would have been produced by an ideal system operating continuously at standard conditions under the same irradiation [7] (Global incident on the plane). The figure 2 shows the performance ratio of case study A.

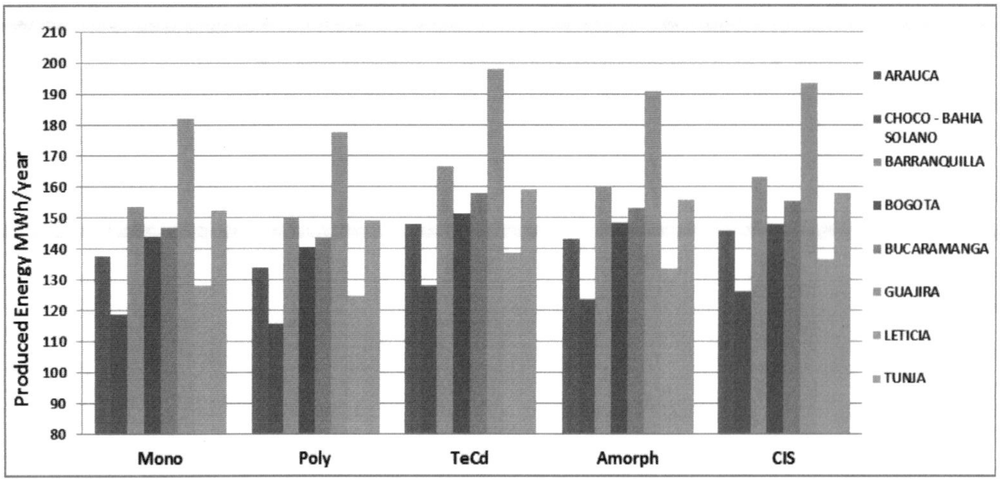

Figure 1. Produced energy of the arrays

3.2 Performance ratio

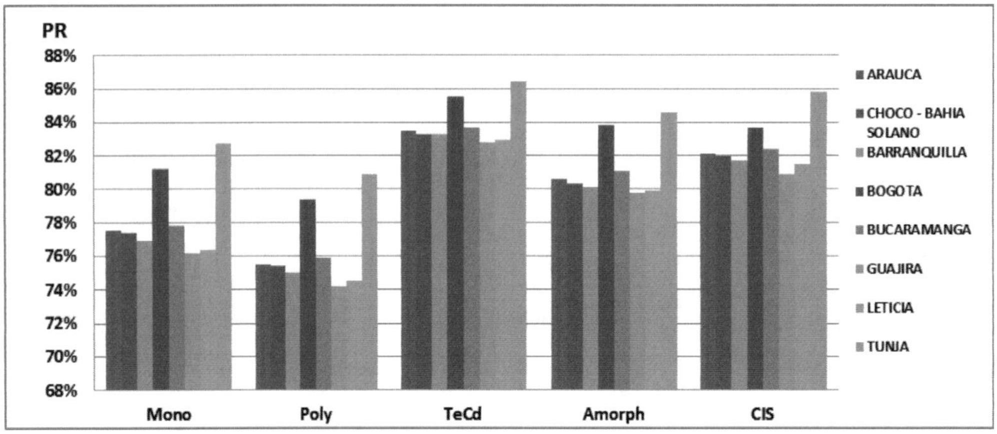

Figure 2. Performance ratio of different arrays

The city with the highest PR is Tunja, followed by Bogotá and Bucaramanga, which are the 3 cities with the lowest average temperatures respectively, so it is safe to say that this parameter is greatly affected by the temperature of the region where the photovoltaic array is installed. The PR increases in places of low temperatures.

In general terms, the technology resulting with higher performance ratio is the Cadmium Telluride type, which explains why it is the technology with the highest production of energy per installed kilowatt, followed closely by two other thin film technologies.

3.3 Temperature losses

The figure 3 shows the comparison of losses by temperature in la Guajira, the region with the highest temperatures, and the opposite case Tunja, with an average temperature of 12.9C.

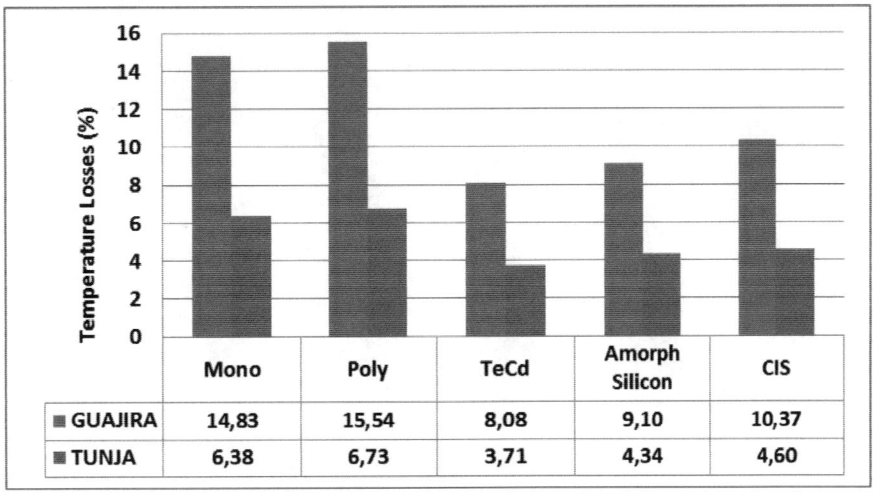

Figure 3. Temperature losses in Guajira and Tunja

4. Case study B: 100kw influenced by wind speed and comparison

In case study B, wind speed has been taken into account as a factor of thermal dispersion of the active surface of the photovoltaic modules.

Even though the wind speed affects the indicators mentioned in the previous case of study (case A), all of them maintain the same relationship between regions and technologies, so it proceeds to a comparative analysis between the 2 cases in order to study only the effects of the wind in the production and the performance ratio (case B). The figure 4 shows the comparison of energy produced of case A and case B in La Guajira.

4.1 Energy production

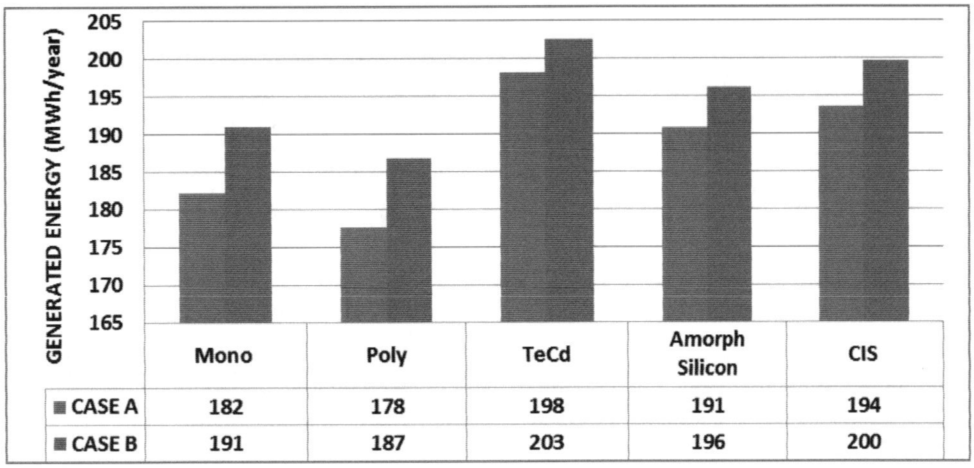

Figure 4. Comparison of studied cases in La Guajira

Taking into account wind speed, the temperature of the surface of the modules is affected in a positive way since photovoltaic panels increase its open circuit voltage as temperature decreases, as seen in figure 5.

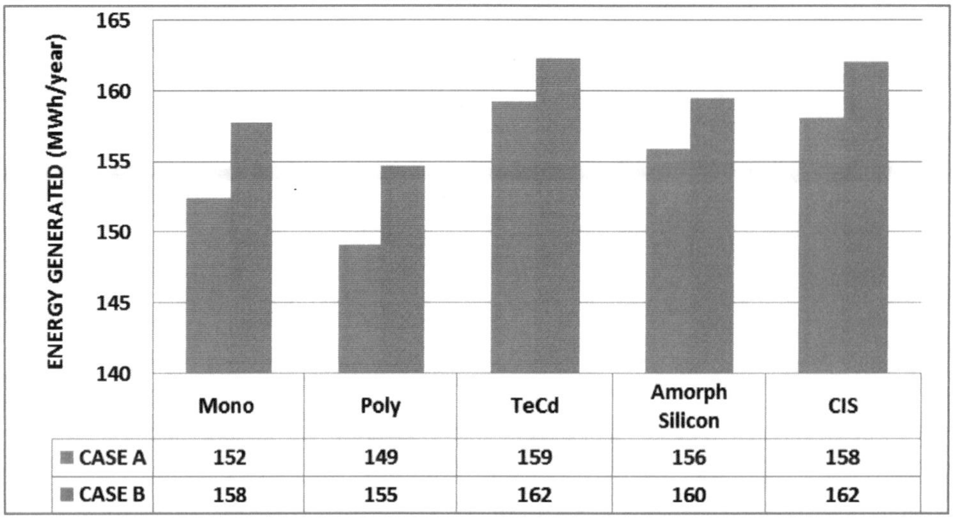

Figure 5. Comparison of studied cases in Tunja

Cities that have higher temperatures presented a higher percentage of increase in energy production because there is higher heat dispersion in trying to reach a thermal equilibrium between the surface of the module [8] and the wind.

4.2 Performance ratio

The figure 6 shows the PR comparison of cases A and B in Bahía Solano city.

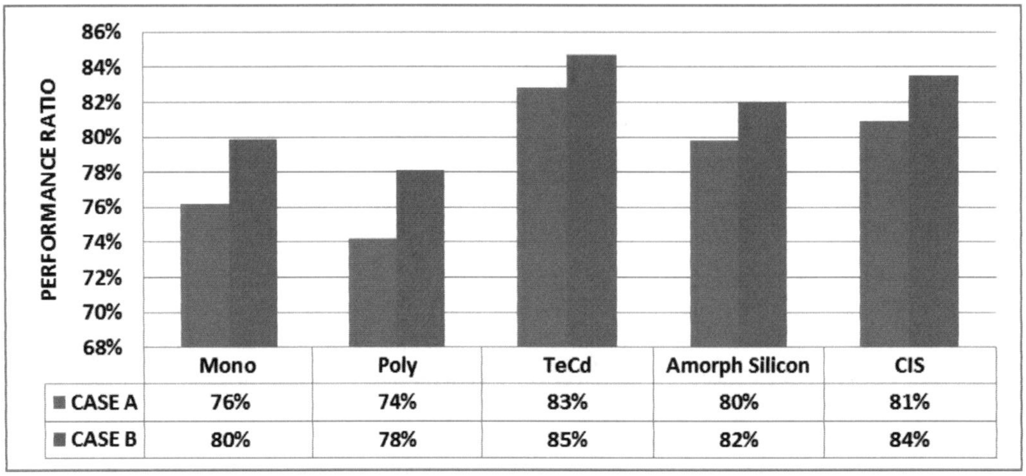

Figure 6. PR comparison in Bahía Solano

The figure 7 shows the PR comparison of cases A and B in La Guajira city. Stands out in particular La Guajira by having the best percentages of increase, due mostly to the high average wind speed in the region, which is around 6.2 m/s.

4.3 Temperature losses

The figures 8 and 9 show the temperature losses in of the arrays in Tunja and La Guajira cities respectively. Analyzing the behavior of the losses by temperature in Tunja and La Guajira, it is clear that the effect of wind reduces these losses, being evident again the importance of the coefficient of temperature [9] in the behavior of photovoltaic modules.

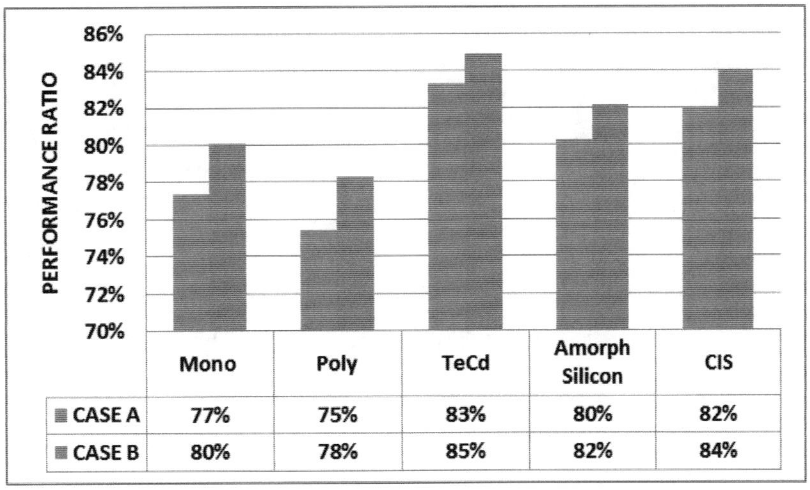

Figure 7. PR comparison in La Guajira

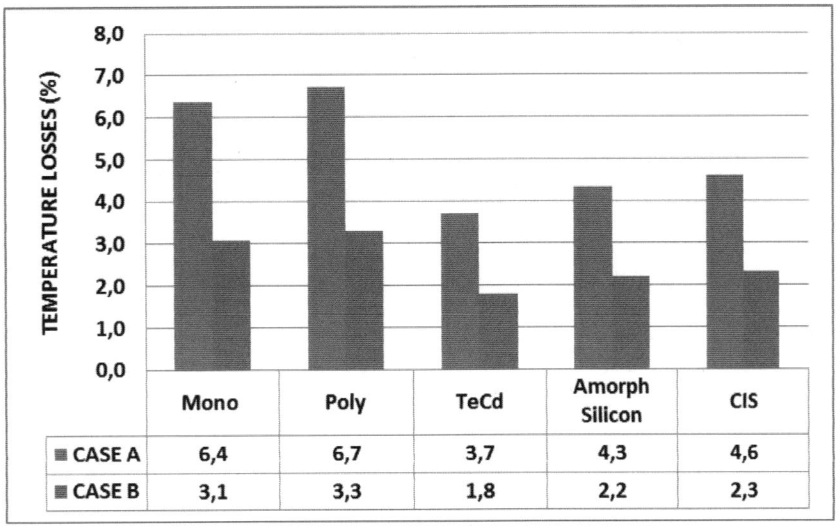

Figure 8. Temperature losses comparison in Tunja

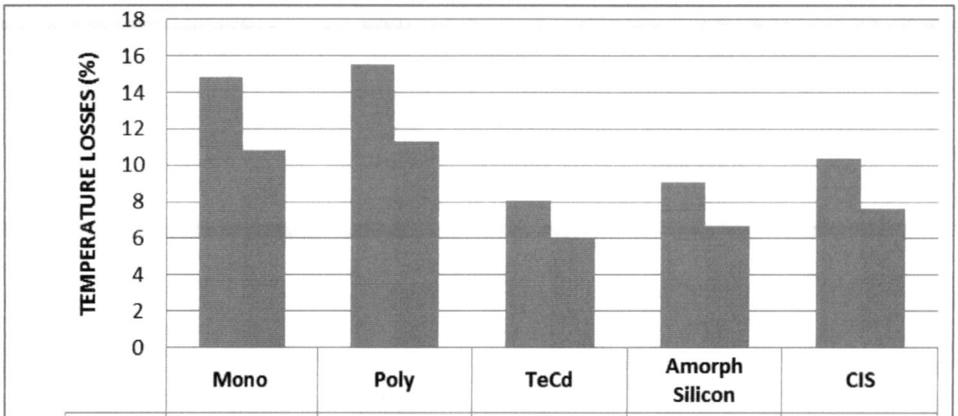

	Mono	Poly	TeCd	Amorph Silicon	CIS
CASE A	14,8	15,5	8,1	9,1	10,4
CASE B	10,8	11,3	6,0	6,7	7,6

Figure 9. Temperature losses comparison in La Guajira

5. Conclusions

It was observed that thin layer technologies generally produce between 3 and 9% more energy than crystalline silicon technologies per installed kilowatt, since they are less affected by temperature which as seen in this document is one of the factors that affects energy production.

In this study was shown the importance of consider the effect of wind speed in the calculation of the performance and energy production of a photovoltaic array. With this consideration, the estimated production rises in average of 3,2% for polycrystalline modules, and 2,2% in thin film modules.

This study has shown, that due to the effect of the temperature in the PV energy production, a PV system in a cold city like Tunja, with an average of 5,03kWh/m^2, can generate a 3,8% more energy per square meter, than the same installation in a warm city with major solar resource of 5,15kWh/m^2, like Bucaramanga.

This analysis has shown that La Guajira is one of the places in Colombia with the greatest photovoltaic potential for photovoltaic energy production, thanks to high levels of radiation and wind speed.

Jungle areas in Colombia, as Leticia and Bahía Solano, are the regions with the worst photovoltaic performance, due to low levels of incident radiation, and high temperatures which meet the necessary conditions so that the photovoltaic arrays show low yields compared with other cities.

References

[1] Lalwani M, Kothari D and Singh M 2010. Investigation of solar photovoltaic simulation softwares *International journal of applied engineering* 585-601.

[2] Petrovic I, Simic Z and Vrazic M 2013. Comparison of pv plant energy generation prediction tools with measured data. *Izlazi u samo elektronickom izdanju* (Nehterlands) 121-124.

[3] Buresch, M. (1983). Photovoltaic energy systems: Design and installation. *New York, McGraw-Hill Book Co,* **347** 1-16.

[4] Delgado J, et al 2009 Measurements and analysis of a 20 kw grid-connected PV-system. *http://e-archivo.uc3m.es/bitstream/handle/10016/5933/PFC_Jorge_Cabeza.pdf?sequence=1*

[5] Mermoud, A. (2012). Pvsyst: Software for the study and simulation of photovoltaic systems. *ISE, University of Geneva, http:// www. pvsyst. com.*

[6] Slamberger J and Virtic P 2013. Determining energy production of the photovoltaic system.

Przeglad Elektrotechniczny.84-87.

[7] Ueda Y, Kurokawa K, Itou T, Kitamura K, Miyamoto Y, Yokota M & Sugihara H. 2006 Performance ratio and yield analysis of grid connected clustered PV systems in Japan. *Photovoltaic Energy Conversion, Conference Record of the 2006 IEEE 4th World Conference on* **2**, 2296-2299.

[8] Virtuani A, Pavanello D, Friesen G 2010 *Overview of temperature coefficients of different thin film photovoltaic technologies* 25th European photovoltaic solar energy conference and exhibition/5th Conference on photovoltaic energy conversion 6-10.

[9] Bertho S, Janssen G, Cleij T, Conings B, Moons W, Gadisa A & Vanderzande D. 2008. Effect of temperature on the morphological and photovoltaic stability of bulk heterojunction polymer: fullerene solar cells. *Solar Energy Materials and Solar Cells*, **92**, 753-760.

AUTHOR INDEX

Institute of Physics
Dirac House, Temple Back
Bristol BS1 6BE UK

CURRAN ASSOCIATES INC
proceedings.com